# Synthetic Plant Growth Regulators

# Synthetic Plant Growth Regulators

*Editor*
*Dr. Dharamvir Hota*

GENE TECH

**2007**

**Gene-Tech Books**

**New Delhi - 110 002**

**ISBN 81-89729-22-5**

***Published by*** : **GENE-TECH BOOKS**
**4762-63/23, Ansari Road, Darya Ganj,**
**NEW DELHI - 110 002**
**Phone: 41562849**
**e-mail: genetechbooks@yahoo.co.in**

***Printed at*** : **Tarun Offset Printers**
**Delhi**

**PRINTED IN INDIA**

# Preface

A plant growth regulator is an organic substance that is made in tiny amounts by the plant and has very definite specific effects on tissue metabolism and growth. The target tissue of the growth regulator may be the local tissue or tissue in a different part of the plant. The growth regulator affects the cell cycle, cell enlargement and cell differentiation. Natural plant growth regulators that move to their target are called plant hormones.

Synthetic plant growth regulators are inorganic substances made by physical chemistry that affect plant growth. Some mimic the natural growth regulators in structure and action. Many are completely different to natural regulators both in structure and action. They are applied for specific purposes to affect specific plant responses. Achieving the best results with PGRs is a combination of art and science - science tempered with a lot of trial and error and a good understanding of plant growth and development. Several new plant growth regulators have come on the market for use in plants in recent years. These plant growth regulators have given plant growers a new opportunity to influence plant growth to their advantage. However, this opportunity goes only as far as the ability of the grower to understand and use them properly.

The structure, functions, advantage and disadvantages of synthetic plant growth regulators elaborately described in this book. Students, teachers and researchers in the field of plant science, agriculture and biotechnology will find this book of utmost use.

***Editor***

# Preface

# Contents

# 1

# Synthetic Plant Growth Regulators

Growth and development of plants is regulated by a combination of genetic factors and environment influences. Hormones can function as signal molecules that trigger signal transduction pathways in cells. Such pathways often result in the synthesis of transcription factors that in turn promote synthesis of enzymes that facilitate chemical reactions within the cell. Signal transduction pathways are equally important for chemical messaging in plants as in animals and both environmental cues and hormones serve as signal stimuli.

Plant hormones, or plant growth regulators, are chemicals produced by plants that alter growth patterns and/or maintenance of the plant. They can be found in many cells and tissues, although plant hormones seem to be concentrated in meristems and buds (which are dormant shoot meristems). Growth regulators control cell activities by sending chemical signals or messages to cells to do something or to not do something, including activating the genes that code for specific enzymes. Plant hormones inhibit as well as promote cellular activities. In contrast to animal hormones, which generally have very specific effects, the hormones identified in plants most often regulate division, elongation and differentiation of cells.

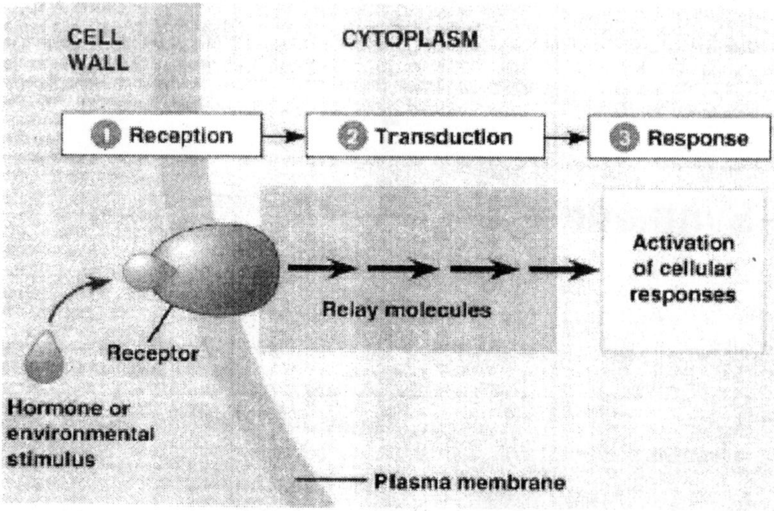


Most hormones have multiple effects in plants. Plant hormones work in very small concentrations. In most cases, the effect plant hormones have on the plant depends on the location of and concentration of the hormone relative to other hormones in the specific tissues. Plant hormones often work in conjunction with each other, and have overlapping effects. They also work with environmental stimuli. There are several classes of plant hormones, including a number of recently "discovered" ones.

### Growth Hormones

Plant hormones, or plant growth regulators, are chemicals produced by plants that alter growth patterns and/or maintenance of the plant. They can be found in many cells and tissues, although plant hormones seem to be concentrated in meristems and buds (which are dormant shoot meristems). Growth regulators control cell activities by sending chemical signals or messages to cells to do something or to not do something, including activating the genes that code for specific enzymes.

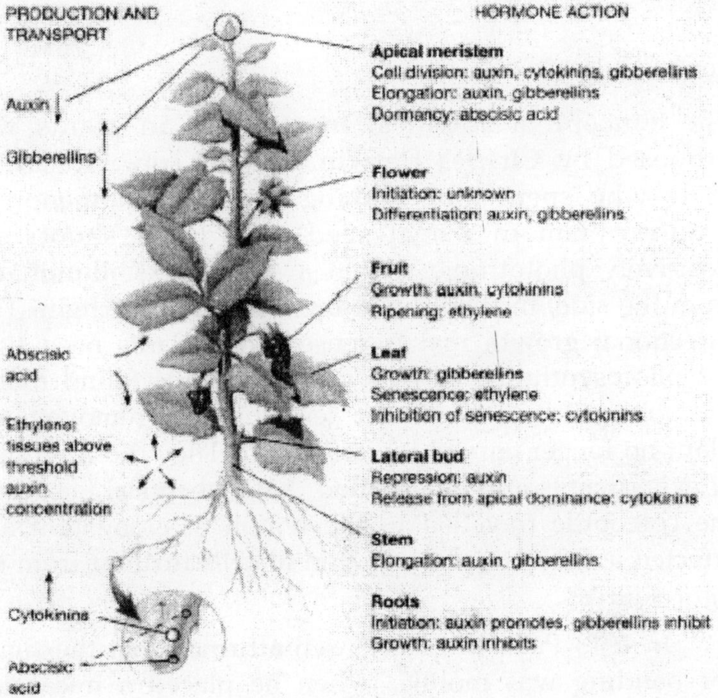


Plant hormones inhibit as well as promote cellular activities. In contrast to animal hormones, which generally have very specific effects, the hormones identified in plants most often regulate division, elongation and differentiation of cells. Most hormones have multiple effects in plants. Plant hormones work in very small concentrations.

In most cases, the effect plant hormones have on the plant depends on the location of and concentration of the hormone relative to other hormones in the specific tissues. Plant hormones often work in conjunction with each other, and have overlapping effects. They also work with environmental stimuli. There are several classes of plant hormones. Each of these are briefly described below.

### *Auxins*

The concept of chemical messengers in plants was proposed by Charles Darwin and his son, Francis, in 1880, who spent time looking at the phenomenon of phototropism in wheat seedlings. Plant shoots are positively phototropic. When a seedling is illuminated from the side, the shoot will bend towards the light. This directional growth makes sense, since plants need light for photosynthesis. However, the Darwins found that if the coleoptile of the wheat seedling was removed, the plant no longer curved towards light. They did a number of experiments and determined that a chemical located in the coleoptile travelled to the region of elongation and effected a differential elongation of cells furthest from the light sources.

In 1913 Peter Boysen-Jensen affirmed that the signal for bending was mobile. When he placed a mica slice between the coleoptile tip and the rest of the shoot, no bending occurred. The mobile chemical was subsequently studied and named auxin by Frits Went in 1926.

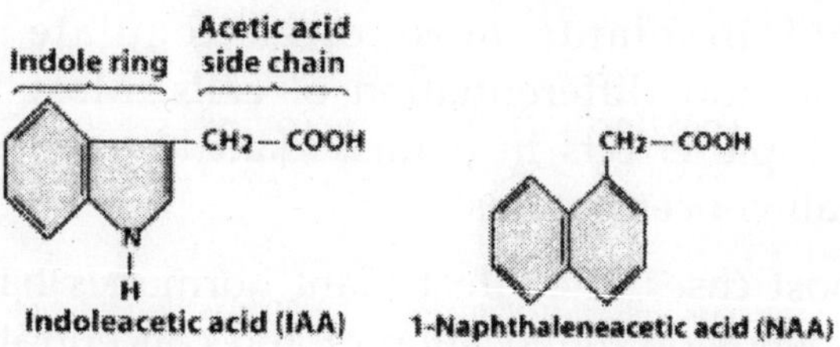

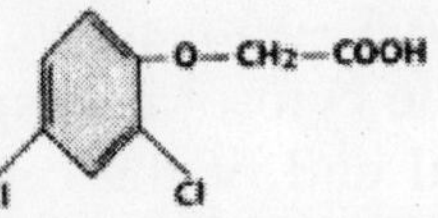

2,4-Dichlorophenoxyacetic acid (2,4-D)

Chemically auxin is indoleacetic acid (IAA), which is synthesised from indole or tryptophan. There are a number of synthetic "auxins" too. Auxins promote growth in molar concentrations of $10^{-3}$ to $10^{-8}$. Auxins are synthesised in shoot apical meristems, young leaves, seeds and fruits.

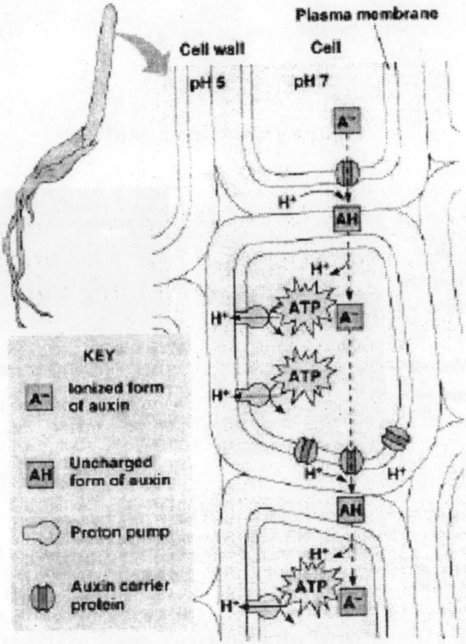


*Auxin Polar Transport*

Auxin always moves down the stem parenchyma cells towards roots by polar (charged) transport (auxin becomes negatively charged) using proton pumps, an

ATP requiring process. Auxin enters cells as IAAH passively, or as IAA—via active cotransport. IAAH dissociates within the cytosol and special auxin transport proteins in the basal end of the cell are needed to carry auxin through the plasma membrane to the top of the adjacent cell. Auxin destined for root tissue, however, moves through phloem sieve tubes.

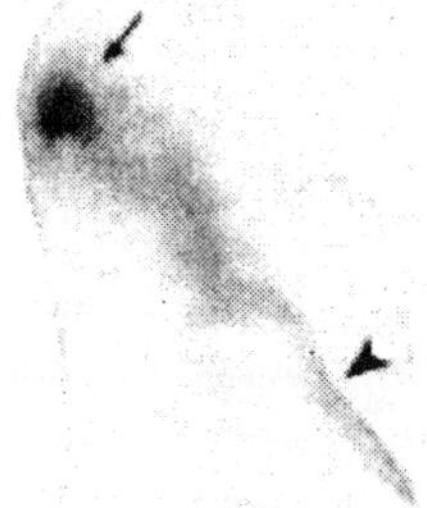

Auxin synthesis in leaf

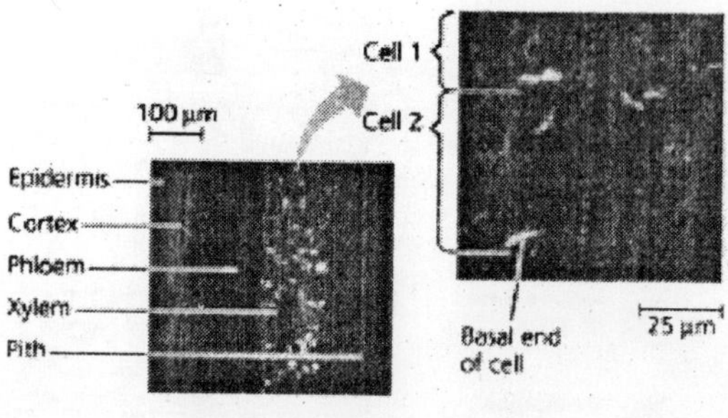


*Auxin Transport Protein*

*Functions of Auxin*

– Auxin promotes elongation and cell enlargement. For cell elongation, proton pumps increase the $H^+$ concentration in cell walls, which stimulates expansins, proteins that disrupt hydrogen bonds and break cross linkages in cellulose. This facilitates wall expansion when cells take in more water.

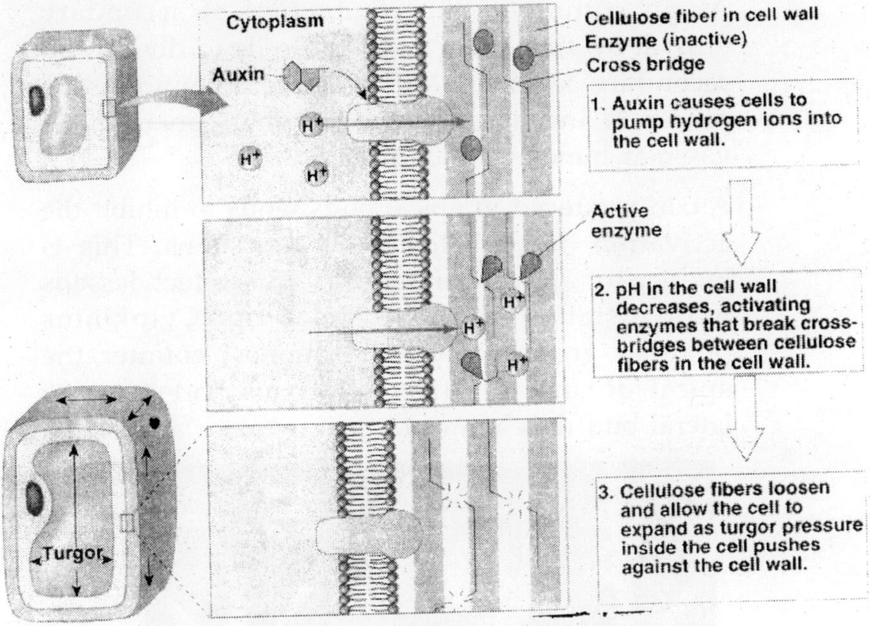


Auxin role in cell wall expansion

– Auxins are involved in tropic responses. Auxins migrate away from light, which accounts for the uneven elongation of cells on the shaded side of a plant unevenly exposed to light, as studied by Darwin.

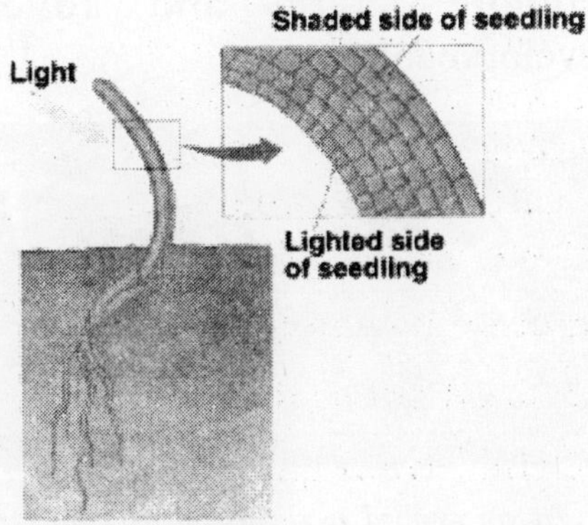

- Auxin stimulates the production of secondary growth by simulating cambium cells to divide and secondary xylem to differentiate. Wound tissue repair is initiated by auxin when portions of vascular bundles are damaged.
- Auxin produced in apical buds tends to inhibit the activation of buds lower on the stems. This is known as apical dominance. This effect lessens with distance from the shoot tip. Cytokinins (another group of plant hormones) counter the apical dominance effect of auxins and promote lateral bud development.

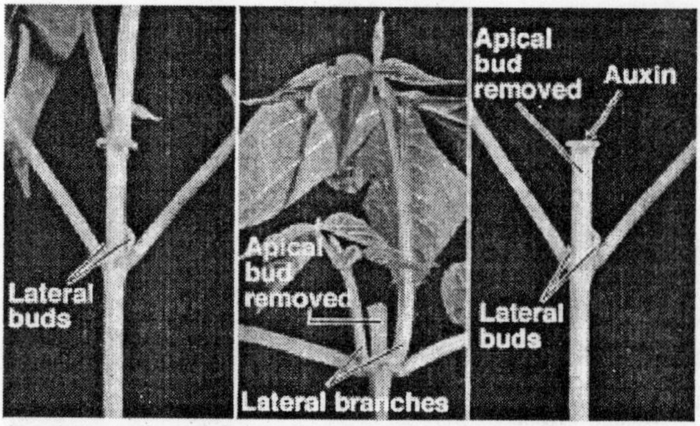


Auxins also:

- Promote lateral and adventitious root development

*Auxin applied to shoots on left and center*

- Promote other hormone production, especially ethylene when auxin concentration increases
- Promote flower initiation
- Loss of auxin → initiates leaf abscission
- Fruit development requires auxin produced by the developing seed. Auxin pastes applied to developing ovaries can promote parthenocarpy. Parthenocarpy is also induced by abscisic acid.

Auxins are toxic in large concentrations. They affect mostly dicots but not monocots. Monocots seem to be able to rapidly degrade the synthetic auxins. Synthetic auxin herbicides include 2-4-D and 2-4-5-T (Agent orange of the Vietnam era contains synthetic auxin)

- Used as defoliants
- Used as weed killers

While "poorly" documented, adverse effects on humans are also asserted from the use of synthetic auxins in high concentrations.

### *Cytokinins*

Cytokinins are a group of phenyl urea derivatives of adenine, one of the molecules in DNA. The first cytokinin was chemically isolated in 1913 and cytokinins were studied using coconut endosperm for a number of years starting in the 1940's. This isolate was shown to be a potent growth promoter and was used in tissue culture and embryo development studies. Folke Skoog and Carlos Miller spent several years attempting to isolate this growth substance. They finally succeeded in isolating a breakdown product of DNA that promoted cell division. They named the substance kinetin and the related group of growth regulators, cytokinins, because they were involved in cell division.

However natural cytokinins were not chemically identified until 1963, when zeatin was identified in corn. Cytokinins are found in actively dividing tissues of seeds, fruits, leaves and root tips, and wound tissue sites. Studies indicate that root tips are most likely the location for cytokinin production and cytokinins are transported through xylem to the rest of the plant. However, localised cytokinins are needed to release buds from dormancy.

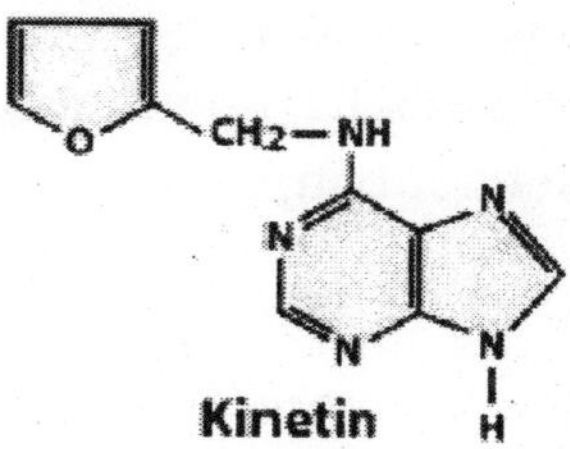
Kinetin

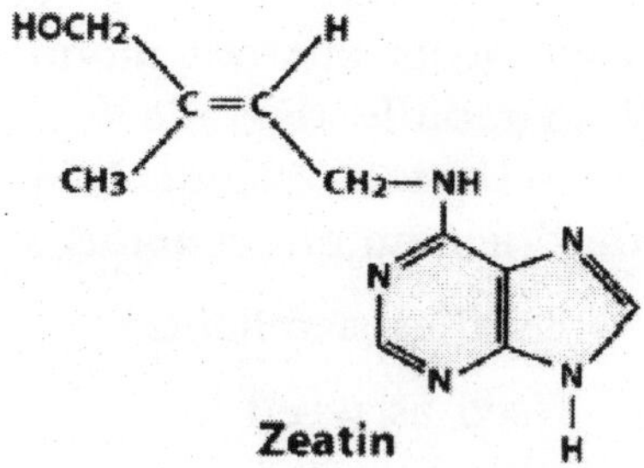
Zeatin

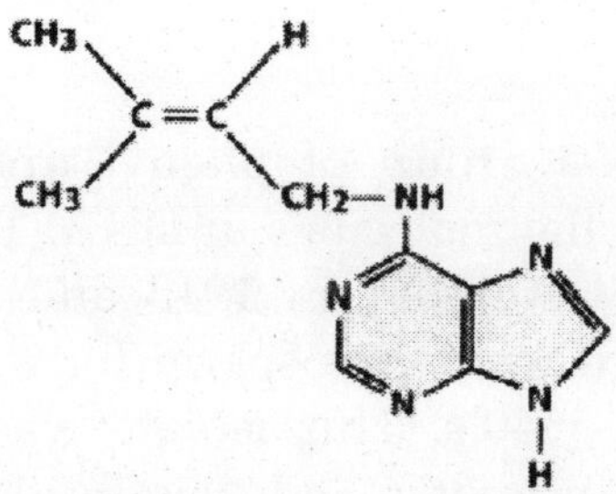
Isopentenyl adenine ($i^6$ Ade)

*Functions of Cytokinin*

– Cytokinins promote cytokinesis. They are found primarily in root meristems, embryos and fruits, and migrate from roots to the shoot systems of plants in xylem tissue.

– The effects of cytokinins are often studied in tissue culture. Normally, cytokinins work in conjunction with auxins, which promote elongation and cell expansion. When auxin is added to a tissue culture medium, cells in the absence of cytokinins enlarge, but division does not take place. When kinetin and IAA are both added to a tissue culture medium, cells divide rapidly producing many small cells. The ratio of cytokinins to auxin controls what tissues differentiate in the tissue culture.

– Cytokinins can promote axillary bud growth by over-riding the apparent inhibiting effect of auxin. This is one of the ways in which plants balance root and shoot growth. Shoot tip auxins inhibit lateral bud activation. Cytokinins produced in root meristems travel upward in xylem and if in higher concentration counter the inhibition of auxins to activate lateral buds.

*Cytokinin treated plant (L)*

*Crown gall tumour*

- Cytokinins also retard leaf senescence, probably by stimulating RNA and protein synthesis and delaying degradation of chlorophyll.
- Some pathogens that promote tumour formation, called galls in plants, take advantage of cytokinin function. The bacteria that form these tumours contain genes for the synthesis of cytokinins so promote rapid undifferentiated cell growth in the infected part of the plant. Agrobacterium tumefaciens, a common vector in plant biotechnology contains such genes.

#### *Gibberellins*

Ewiti Kurosawa is credited with the discovery of gibberellins when he determined that a fungus was responsible for abnormal rice seedling growth, called the "foolish seedling" disease. The fungus secreted a chemical that caused the rice plants to grew abnormally long, and then collapse from weakness. The fungus was Gibberella *fujikuroi*, hence the hormone name.

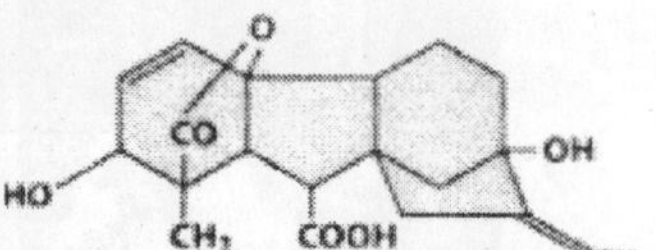


Gibberellic acid $(GA_3)$

*Normal and foolish rice*

Many seeds contain a variety of different gibberellins. Over 100 different gibberellins (organic acids synthesised from mevalonic acid) are known. Gibberellins are produced in roots and younger leaves, but have their highest concentration in seeds. Most effects of gibberellins are shown only in concert with auxins.

*Functions of Gibberellin*

- Gibberellins work with auxins to promote rapid elongation and division of stem tissue. Gibberellins determine microtubule alignment in the preprophase band that determines the plane of cellulose expansion.

The effects of gibberellins on elongation are seen in:

- Bolting of biennials, to produce flowers during the first growing season
- Reversal of genetic dwarfism

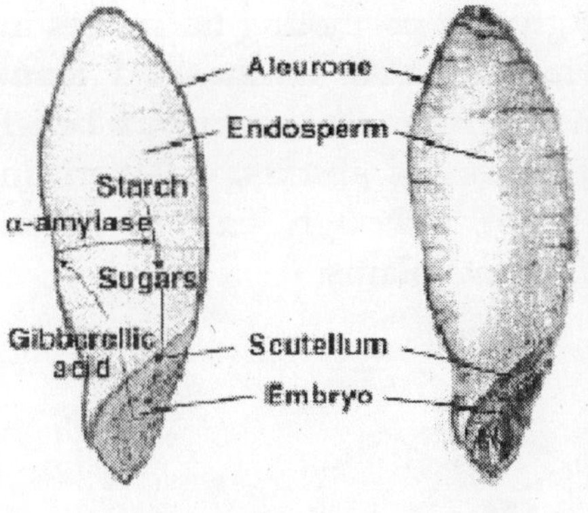


*Gibberellin mobilising nutrients*

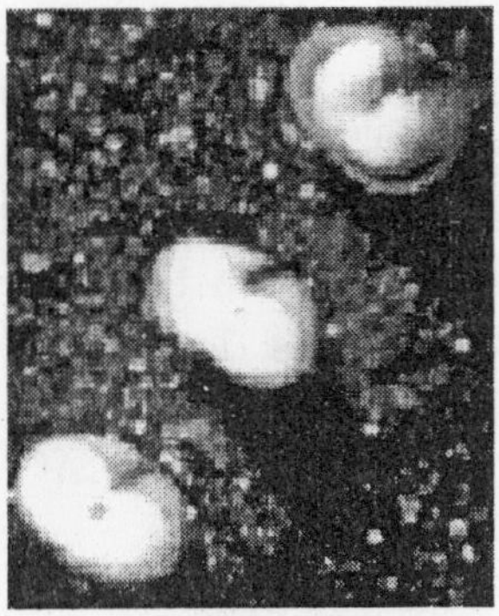

*Sugar production in seed*

- Gibberellins are important to breaking dormancy after imbibition of water by the seed coat. Gibberellins signal germination activities. In particular, application of gibberellins to seeds will counter the normal environmental cues, such as exposure to low temperatures. They stimulate RNA to promote synthesis of enzymes that convert stored nutrients (starches) to sugars needed for rapid cell respiration during germination. Abscisic acid can counter the effect of gibberllin to keep seeds dormant.

– Gibberellins can, as can auxins, promote parthenocarpy in many plants, which has commercial applications, although the most significant use in the grape industry, where grapes grow larger, and with longer internodes.

***Abscisic Acid (ABA)***

Abscisic acid (ABA) is a hormone that functions by inhibiting growth activities in times of environmental stress rather than by promoting growth. It often serves as an antagonist to the other growth promoting hormones in plants. Abscisic acid, which is also synthesised from mevalonic acid, got its name from the erroneous belief that it promoted the formation of abscission layers in leaves and fruits. It does not, although leaf abscission accompanies dormancy in many plants.

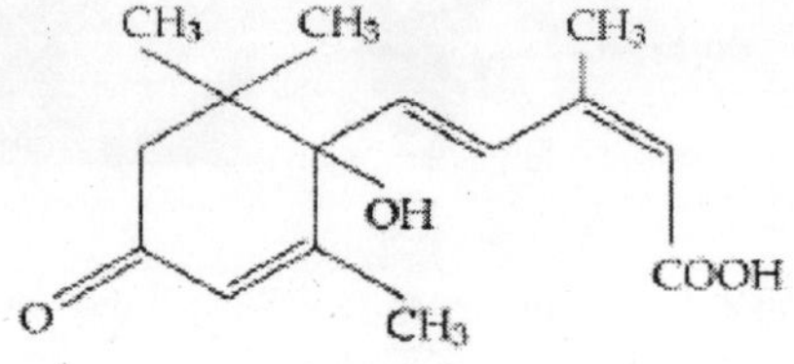

Abscisic acid

*Functions of Abscisic acid*

– ABA promotes seed dormancy activities. ABA levels are high when seeds mature, promoting lowered metabolism and synthesis of proteins needed to withstand the dehydration associated with dormancy. Seeds germinate when ABA is degraded by some environmental action. Desert seeds must have the ABA washed out of the seed coat, as do plants of marshy areas; temperate area plants have ABA degraded by light-stimulated enzymes. In other cases breaking dormancy is relative to the ratio of ABA (which keeps seeds in dormancy) and gibberellins (which promote germination).

Low levels of ABA in maturing seeds promote premature germination.

- Similar to seed dormancy, ABA promotes winter bud scale formation on woody plants in preparation for winter dormancy. ABA derivatives, called dormins, are used in commercial nurseries to keep materials to be shipped in dormant conditions. The dormancy can be reversed with gibberellins.
- ABA is also referred to as the stress activity hormone. For example, ABA promotes stomata closure during leaf water deficit conditions by activating K+ ion transport out guard cells. This involves signal transduction pathways with Calcium secondary messengers. ABA in this case originates in roots, and detects low water level in root tissues. ABA moves upward into leaves and activates stomatal closure.

#### Ethylene

Ethylene is the sole plant growth regulator known that is a gas. (It is a small hydrocarbon molecule, easily synthesised in chemistry labs). You will recall that nitric oxide is involved in signal transduction pathways leading to smooth muscle contraction.

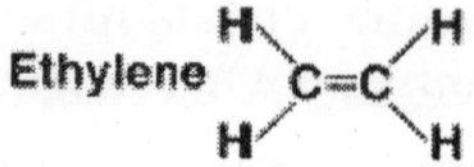


Ethylene is produced in many plant organs, though its effects are most studied in fruits. Ethylene is the signal molecule for signal transduction pathways that activate a number of tissue growth and development activities throughout the plant, including fruit maturation, leaf abscission and senescence.

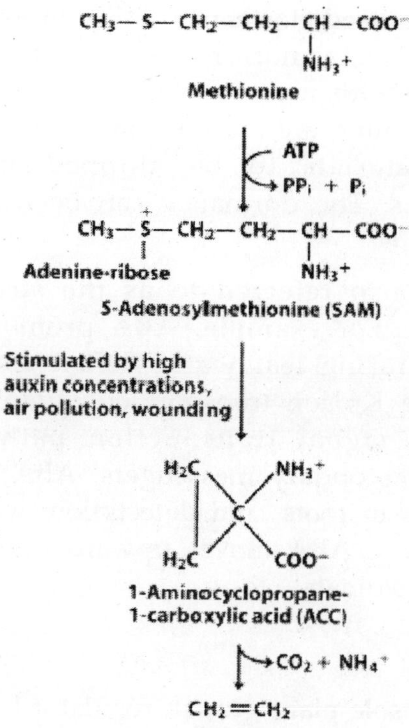

Ethylene is synthesised from the amino acid methionine. High auxin concentrations promote an intermediate step in this pathway that activates an enzyme in the tonoplast (vacuole membrane) to convert the intermediate into ethylene. Toxic substances such as air pollutants also trigger the intermediate step. Ethylene is also produced in wounded or bruised tissues. The benefit of this is unknown; it may be an artifact of the wounding response.

#### *Functions of Ethylene*

The benefits of ethylene as a fruit ripener were known for centuries prior to it being identified as a plant product in the early 1900's. Chinese gardeners knew centuries ago that fruits ripened better in rooms with burning incense. Citrus growers used kerosene stoves in the rooms in which they ripened their fruit. During the era of gas lamps, leaking lamps along city streets often promoted leaf abscission.

Today, grocer warehouses have ethylene rooms that are used for ripening most of our produce, which is shipped unripe. Immature fruits are firmer and less subject to damage. Ethylene activates ripening in fruits by signalling chemical reactions that degrade the pectins of the middle lamella, softening fruit, and promoting the conversion of stored starches and/or oils into sugars that attract seed dispersers. Some fruit ripening involves a significant increase in the rate of cell respiration with concurrent high $O_2$ uptake. This function of ethylene (fruit ripening) has extensive agricultural impact.

Biotechnology has been used to alter sensitivity to ethylene in some crops to facilitate harvest and shipping of fruits. Using gene technology, tomatoes lacking ethylene receptor genes have been grown and their fruits have a shelf life of 100 days!

- Ethylene promotes female flower production in some members of the Cucurbitaceae, whereas high gibberellins may promote formation of male flowers.
- Ethylene functions to help germinating seeds overcome mechanical stress. A shoot tip that encounters an immovable object will grow around the object by changing its growth direction. This occurs through differential elongation of cell walls. When the shoot tip cannot penetrate the mechanical obstacle, ethylene is synthesised, which slows cell wall expansion. The walls thicken and the stem curves to bypass the obstacle via horizontal growth. The stem tip pushes upward periodically to test for the obstacle. If the obstacle is encountered, ethylene initiates another cycle of stem thickening and horizontal growth. If no obstacle is encountered, the shoot resumes negative geotropic growth.

This growth response is known as the triple response (decrease in elongation, stem thickening, and change in orientation of shoot growth to horizontal).

- Ethylene is responsible for initiating the programmed death (apoptosis) of sclerenchyma and xylem vessels and tracheids. Death involves intense cellular activity to degrade and salvage materials of the cytoplasm.
- Ethylene is the direct cause of leaf and fruit abscission. Ethylene promotes the degradation of the cell walls in the abscission zone cells. The declining levels of auxin in leaves signal the production of ethylene in the abscission zone cells as summer ends.

The auxin/ethylene interaction for abscission is also used commercially. Fruit growers may spray auxins on fruits to prevent the fruits from falling to the ground prior to harvest. Concentration is important; very high auxin level promotes ethylene production. High concentrations of $CO_2$ inhibit ethylene production so $CO_2$ can be used to prevent fruit ripening in grocer warehouses.

### *Brassinolides*

Brassinolides is a plant steroid discovered in pollen of members of the mustard family, and studied in *Arabadopsis*. Chemically the brassinosteroids are similar to animal steroid hormones, and have similar genes that code for the steroid synthesis pathways. The functions of brassinosteorids overlap those of other plant hormones, and may be additive. Brassinosteoroids activate signal transduction pathways that promote cell elongation and cell division. Brassinosteroids also promote differentiation of xylem tissue, and perhaps other tissues, too. Brassinosteroids can also retard leaf abscission. Absence of brassinolides results in dwarf plants. It is difficult to study the brassinosteroids because their effects overlap those of auxin and gibberellins.

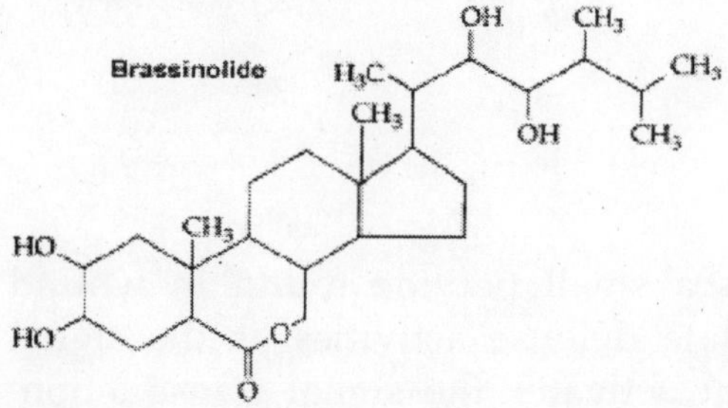


### *Salicylic Acid*

Salicylic acid is known to activate defence genes for

resistance against pathogen invaders, known as the hypersensitive response. Salicylic acid, a phenolic extract from willow bark, was long used as an analgesic. It is now prepared commercially and is the active ingredient of aspirin.

Salicylic acid

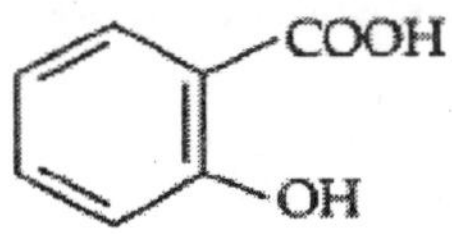


*Oligosaccharins*

Oligosaccharins are short chain sugars in cell walls that may have a role in defense against pathogens. When a pathogen "chomps" on the oligosaccharide, the oligosaccharide activates a signal transduction pathway leading to plant defense responses. They may also help regulate growth, differentiation and flower development all by activating signal transduction pathways.

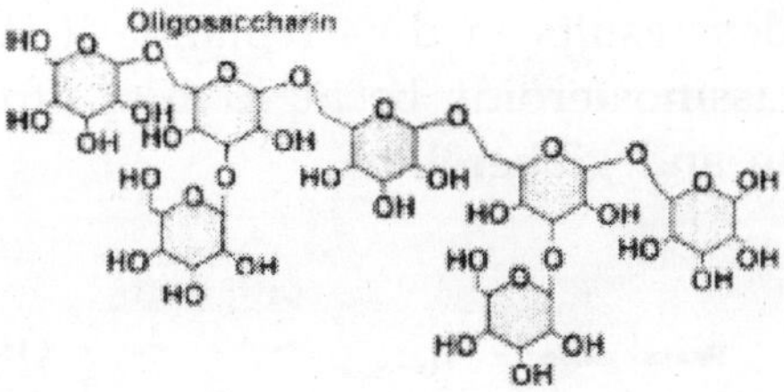


*Systemin*

Systemin is a small peptide found in wound tissue. It may stimulate defense activities as the signal molecule (ligand) that activates the signal transduction pathways that include the jasmonates.

***Jasmonates***

Jasmonates are a group of fatty acid derivatives. They appear to have a role in seed germination, root growth, and the storage of protein (especially in seeds).

Jasmonic acid

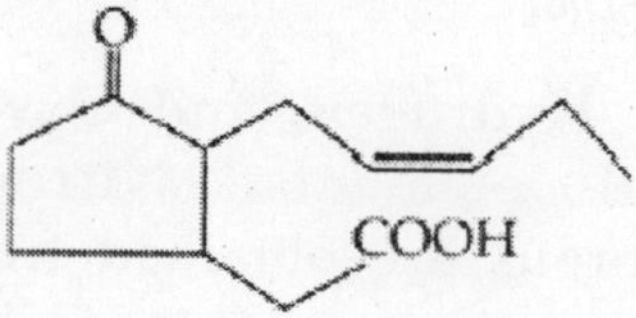

Jasmonates synthesised in response to signal molecules produced in wound areas are involved in the signal transduction pathways that result in secondary metabolites (protease inhibitors) that poison the predator, or synthesis of volatile molecules that attract the predator's predators.

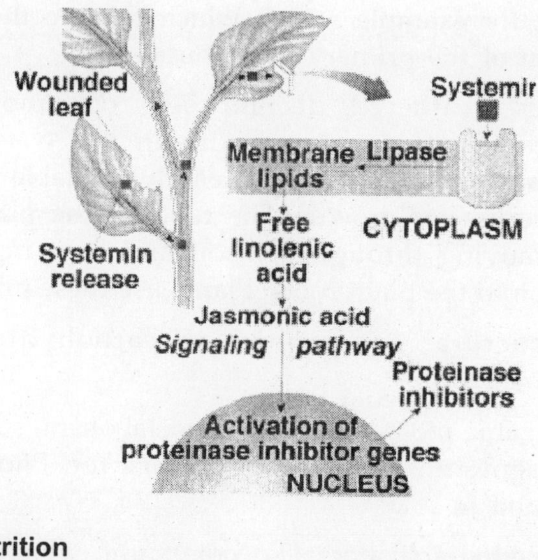


**Plant Nutrition**

Unlike animals plants obtain their nutrition from the soil and atmosphere. Using sunlight as an energy source, plants are capable of making all the organic macromolecules they need by modifications of the sugars they form by photosynthesis. However, plants must take up various minerals through their root systems for use.

### Balanced Diet

Carbon, Hydrogen, and Oxygen are considered the essential elements. Nitrogen, Potassium, and Phosphorous are obtained from the soil and are the primary macronutrients. Calcium, Magnesium, and Sulphur are the secondary macronutrients needed in lesser quantity. The micronutrients, needed in very small quantities and toxic in large quantities, include Iron, Manganese, Copper, Zinc, Boron, and Chlorine. A complete fertiliser provides all three primary macronutrients and some of the secondary and micronutrients. The label of the fertiliser will list numbers, for example 5-10-5, which refer to the percent by weight of the primary macronutrients.

Soil is weathered, decomposed rock and mineral (geological) fragments mixed with air and water. Fertile soil contains the nutrients in a readily available form that plants require for growth. The roots of the plant act as miners moving through the soil and bringing needed minerals into the plant roots. Plants use these minerals in:

1. Structural components in carbohydrates and proteins
2. Organic molecules used in metabolism, such as the Magnesium in chlorophyll and the Phosphorous found in ATP
3. Enzyme activators like potassium, which activates possibly fifty enzymes
4. Maintaining osmotic balance

### *Mycorrhizae, Bacteria, and Minerals*

Plants need nitrogen for many important biological molecules including nucleotides and proteins. However, the nitrogen in the atmosphere is not in a form that plants can utilise. Many plants have a symbiotic relationship with bacteria growing in their roots: organic nitrogen as rent for space to live. These plants tend to have root nodules in which the nitrogen-fixing bacteria live.

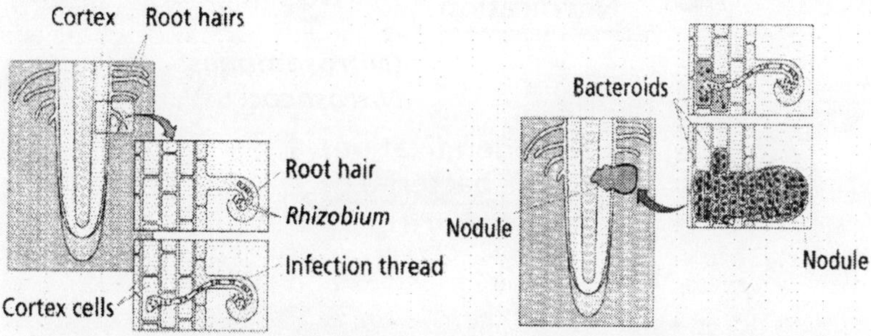


*Development of a root nodule, a place in the roots of certain plants, most notably legumes (the pea family), where bacteria live symbiotically with the plant.*

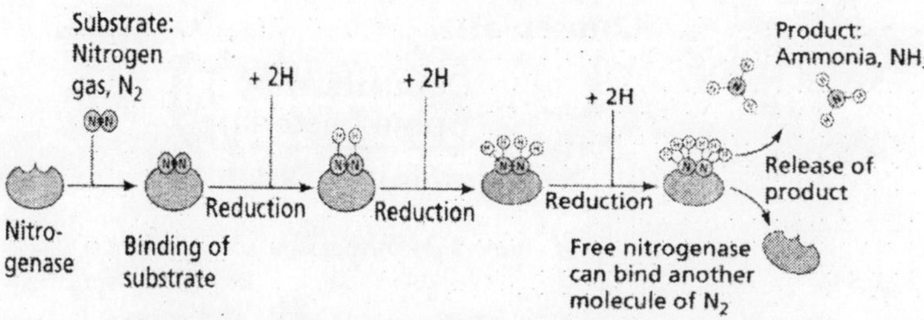


*Pathway for converting (fixing) atmospheric nitrogen, $N_2$, into organic nitrogen, $NH_3$.*

All the nitrogen in living systems was at one time processed by these bacteria, who took atmospheric nitrogen ($N_2$) and modified it to a form that living things could utilise.

Not all bacteria utilise the above route of nitrogen fixation. Many that live free in the soil, utilise other chemical pathways.

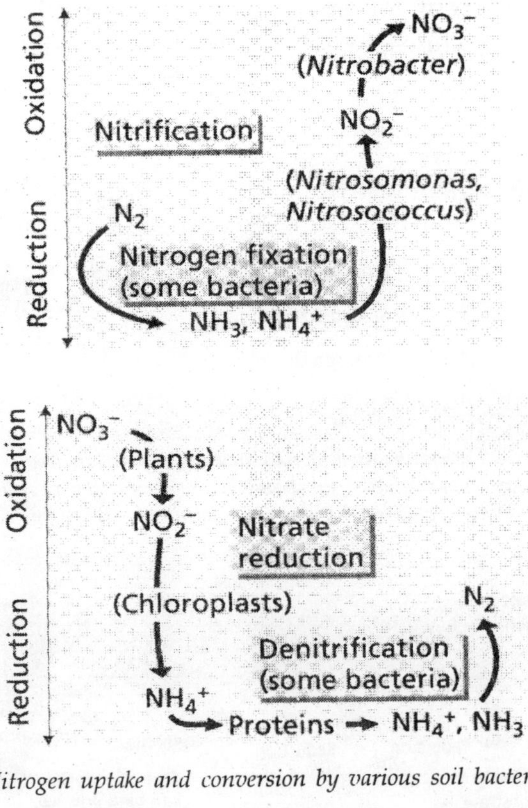


*Nitrogen uptake and conversion by various soil bacteria.*

Roots have extensions of the root epidemal cells known as root hairs. While root hairs greatly enhance the surface area (hence absorbtion surface), the addition of symbiotic mycorrhizae fungi vastly increases the area of the root for absorbing water and minerals from the soil.

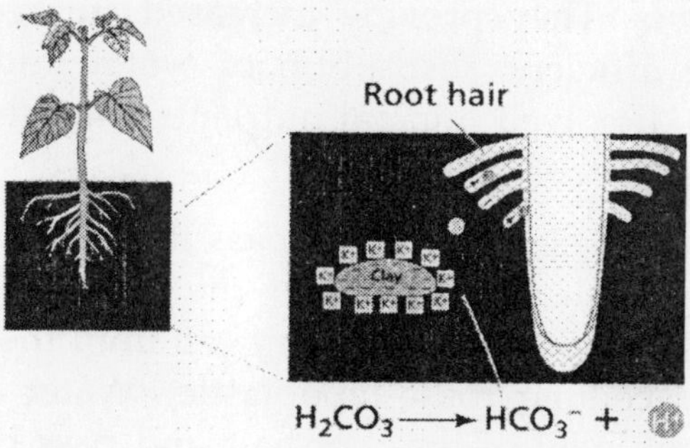


*Role of the root hairs in increasing the surface area of roots to promote increased uptake of water and minerals from the soil.*

## Absorption of Water and Mineral

Animals have a circulatory system that transports fluids, chemicals, and nutrients around within the animal body. Some plants have an analogous system: the vascular system in vascular plants; trumpet hyphae in bryophytes.

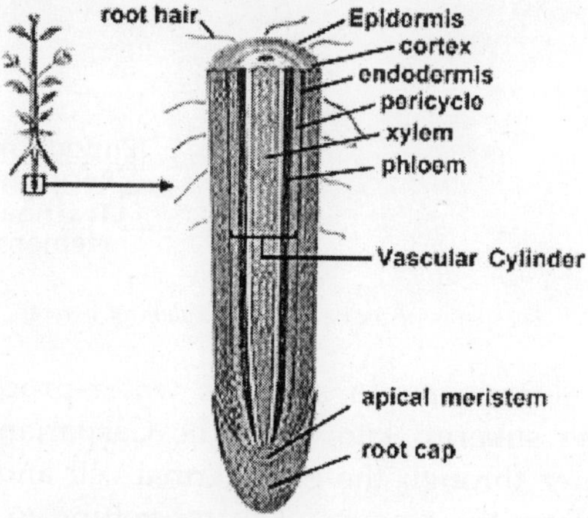


*Structure of a typical root.*

Root hairs are thin-walled extensions of the epidermal cells in roots. They provide increased surface area and thus more efficient absorption of water and minerals. Water and dissolved mineral nutrients enter the plant via two routes.

Water and selected solutes pass through only the cell membrane of the epidermis of the root hair and then through plasmodesmata on every cell until they reach the xylem: intracellular route (apoplastic). Water and solutes enter the cell wall of the root hair and pass between the wall and plasma membrane until the encounter the endodermis, a layer of cells that they must pass through to enter the xylem: extracellular route (symplastic).

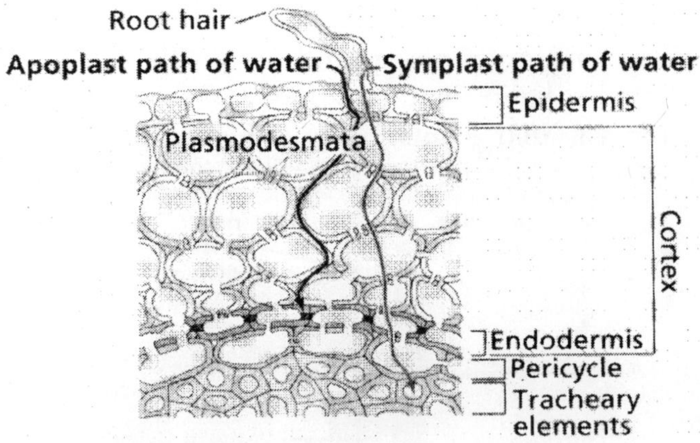


*The paths of water into the xylem of a root.*

The endodermis has a strip of water-proof material (containing suberin) known as the Casparian strip that forces water through the endodermal cell and in such a way regulates the amount of water getting to the xylem.

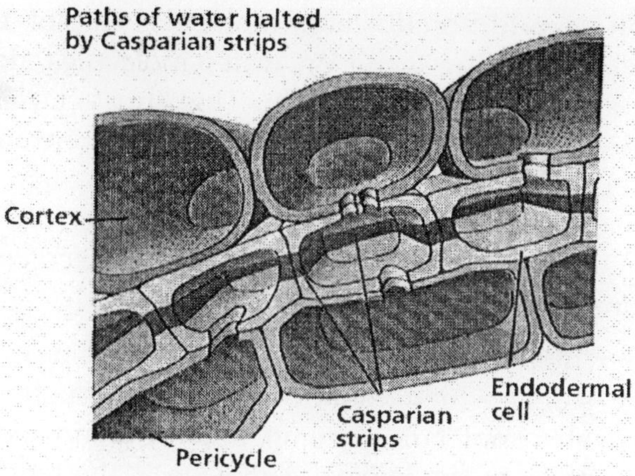


*Casparian strip.*

Only when water concentrations inside the endodermal cell fall below that of the cortex parenchyma cells does water flow into the endodermis and on into the xylem.

### *Xylem*

Xylem is the water transporting tissue in plants that is dead when it reaches functional maturity. Tracheids are long, tapered cells of xylem that have end plates on the cells that contain a great many crossbars. Tracheid walls are festooned with pits. Vessels, an improved form of tracheid, have no (or very few) obstructions (crossbars) on the top or bottom of the cell. The functional diameter of vessels is greater than that of tracheids. Water is pulled up the xylem by the force of transpiration, water loss from leaves.

Mature corn plants can each transpire four gallons of water per week. Transpiration rates in arid-region plants can be even higher. Water molecules are hydrogen bonded to each other. Water lost from the leaves causes diffusion of additional water molecules out of the leaf vein xylem, creating a tug on water molecules along the water columns within the xylem. This "tug" causes water molecules to rise up from the roots to eventually the leaves. The loss of water from the root xylem allows additional water to pass throught the endodermis into the root xylem. Cohesion is the ability of molecules of the same kind to stick together.

Water molecules are polar, having slight positive and negative sides, which causes their cohesion. Inside the xylem, water molecules are in a long chain extending from the roots to the leaves. Adhesion is the tendency of molecules of different kinds to stick together. Water sticks to the cellulose molecules in the walls of the xylem, counteracting the force of gravity and aiding the rise of water within the xylem.

### Cohesion-Adhesion Theory

Transpiration exerts a pull on the water column within the xylem. The lost water molecules are replaced by water from the xylem of the leaf veins, causing a tug on water in the xylem. Adhesion of water to the cell walls of the xylem facilitates movement of water upward within the xylem. This combination of cohesive and adhesive forces is referred to as the Cohesion-Adhesion Theory.

### Guard Cells

In most environments, the water concentration outside the leaf is less than that inside the leaf, causing a loss of water through openings in the leaf known as stomata (singular = stoma). Guard cells are crescent-shaped cells of the epidermis that flank the stoma and regulate the size of the opening. Together, the guard cells and stoma comprise the stomatal apparatus. The inner wall of the guard cell is thicker than the rest of the wall. When a guard cell takes up potassium ions, water moves into the cell, causing the cell to become turgid and swell, opening the stoma.

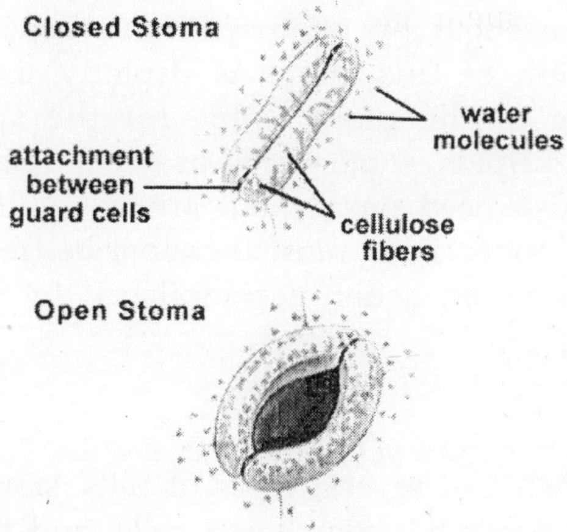


*Functioning of guard cells in the opening and closing of stomata.*

When the potassium leaves the guard cell, the water also leaves, causing plasmolysis of the cells, and a closing of the stoma.

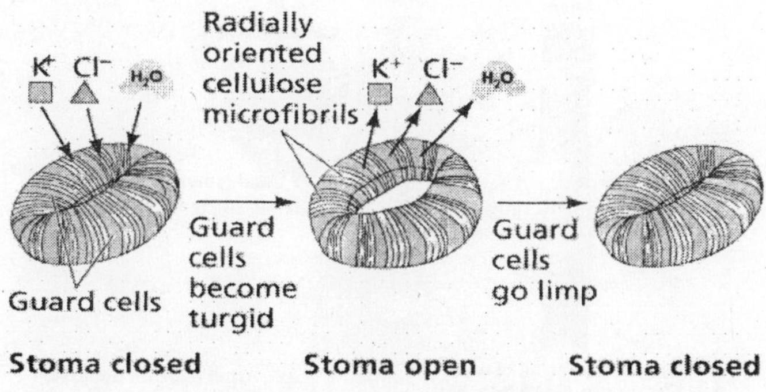


*Ions and stomatal function.*

Stomata occupy 1% of the leaf surface, but account for 90% of the water lost in transpiration.

### *Nutrient Storage*

Plants make sugar by photosynthesis, usually in their leaves. Some of this sugar is directly used for the metabolism of the plant, some for the synthesis of proteins and lipids, some stored as starch. Other parts of the plant also need energy but are not photosynthetic, such as the roots. Food must therefore be transported in from a source, an action accomplished by the phloem tissue.

### *Phloem*

Phloem consists of several types of cells: sieve tube cells (aka sieve elements), companion cells, and the vascular parenchyma. Sieve cells are tubular cells with endwalls known as sieve plates. Most lose their nuclei but remain alive, leaving an empty cell with a functioning plasma membrane. Companion cells load sugar into the sieve element (sieve elements are connected into sieve tubes).

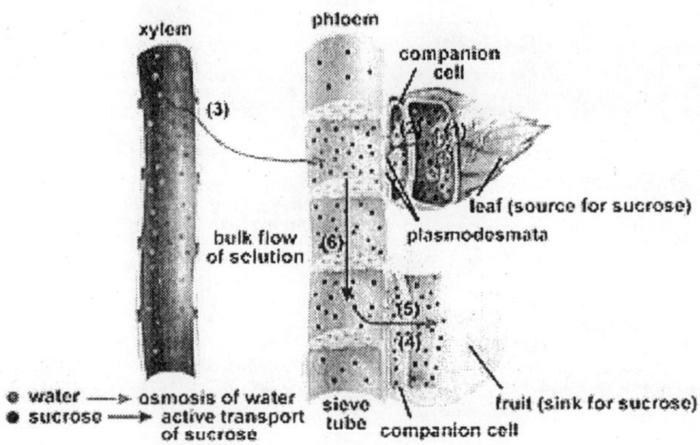


*Bulk flow and the movement of food from sources to sinks.*

Fluids can move up or down within the phloem, and are translocated from one place to another. Sources are places where sugars are being produced. Sinks are places where sugar is being consumed or stored. Food moves through the phloem by a Pressure-Flow Mechanism. Sugar moves (by an energy-requiring step) from a source (usually leaves) to a sink (usually roots) by osmotic pressure.

Translocation of sugar into a sieve element causes water to enter that cell, increasing the pressure of the sugar/water mix (phloem sap). The pressure causes the sap to flow toward an area of lower pressure, the sink. In the sink, the sugar is removed from the phloem by another energy-requiring step and usually converted into starch or metabolised.

#### *External Stimulus*

One plant response to environmental stimulus involves plant parts moving toward or away from the stimulus, a movement known as a tropism. Nastic movements are plant movements independent of the direction of the stimulus.

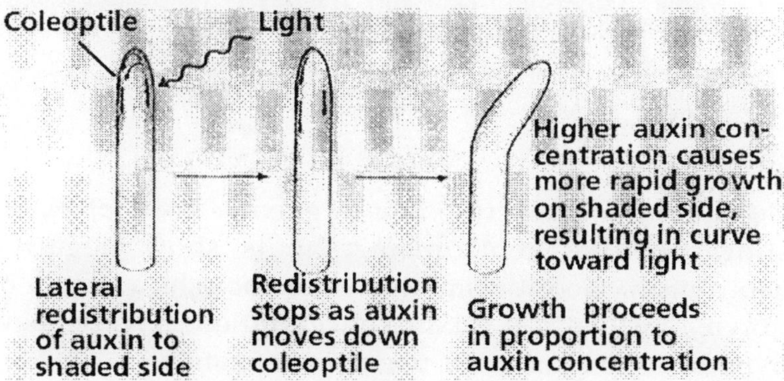


*Phototropism in the coleoptile of a monocot.*

Charles Darwin and his son Francis studied the familiar reaction of plants growing toward light: phototropism. The Darwins discovered that the tips of the plant curved first, and that the curve extended gradually down the stem. By covering the tips with foil, they prevented the plant from curving. They concluded that some factor was transmitted from the tip of the plant to the lower regions, causing the plant to bend.

We now know, from the 1926 experiments of Frits Went, that auxin, a plant hormone produced in the stem tip (auxins promote cell elongation), moves to the darker side of the plant, causing the cells there to grow larger than corresponding cells on the lighter side of the plant. This produces a curving of the plant stem tip toward the light, a plant movement known as phototropism.

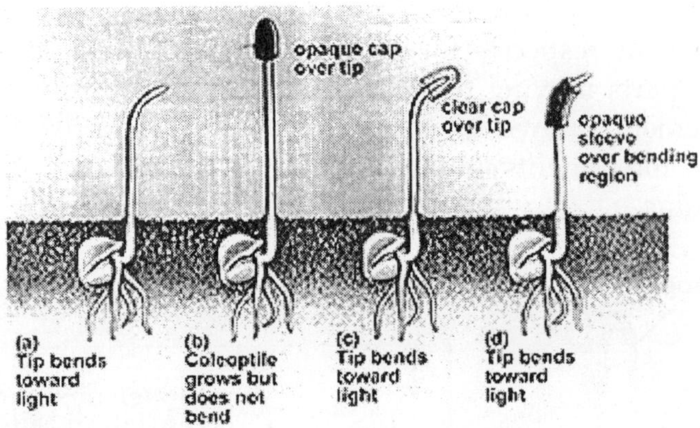


Geotropism is plant response to gravity. Roots of plants show positive geotropism, shoots show negative geotropism. Geotropism was once thought a result of gravity influencing auxin concentration. Several new hypotheses are currently under investigation.

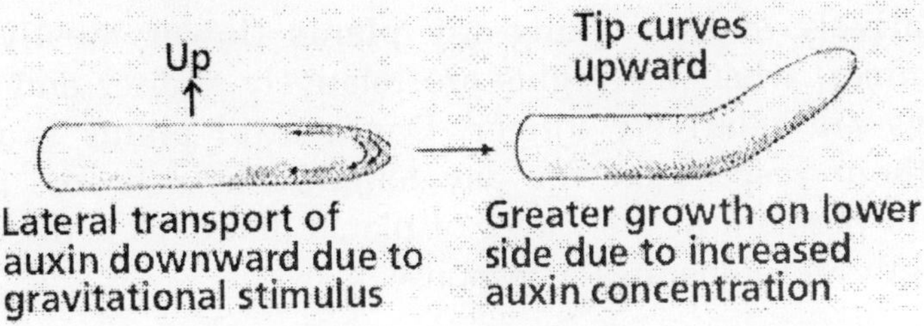


*Geotropism.*

Thigmotropism is plant response to contact with a solid object. Tendrils of vines warp around objects, allowing the vine to grow upward.

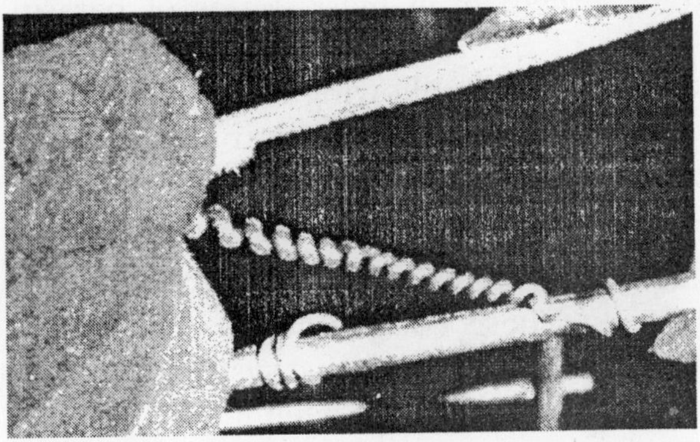

*Curling of a tenbdril around a metal support, an example of thigmotropism.*

Nastic movements, such as nyctinasty, result from several types of stimuli, including light and touch. Legumes turn their leaves in response to day/night conditions. Mimosa, also known as the sensitive plant, has its leaves close up when touched. Photoperiodism is the plant response to the relative amounts of light and dark in a 24 hour period, and controls the flowering of many plants.

Short-day plants flower during early spring or fall, when the nights are relatively longer and the days are relatively shorter. Long-day plants flower mostly in summer, when the nights are relatively shorter and the days are relatively longer. Day-neutral plants flower without respect for the day length. Phytochrome is a plant pigment in the leaves of plants that detects the day length and generates a response.

### *Plant Secondary Compounds*

Plants produce primary compounds important in their metabolism. They also produce secondary compounds that serve to attract pollinators, kill parasites, and prevent infectious diseases. Pea plants produce pisatin, a chemical that protects them from most strains of parasitic fungi. Some strains of the fungus (Fusarium) contain enzymes that inactivate pisatin, allowing them to infect pea plants. Some plants produce natural insecticides, such as pyrethrum, a chemical produced by chrysanthemums that is also commercially available to gardeners.

Antinutrients are chemical produced when plants are under attack. These compounds inhibit the action of enzymes in the insect's digestive system. More that 10,000 defensive chemicals have been identified, including caffeine, phenol, tannin, nicotine, and morphine. Some plant secondary compounds are useful to humans as

1. pesticides
2. medicines (salicylic acid, the main component in aspirin)
3. stimulants (caffeine, the most widely used psychoactive drug in the world)
4. chewing gum (chicle, a compound from the sapodilla tree in Mexico was used in the first chewing gum).

# 2

# Photosynthesis to Stress and Plant Growth Regulators

In green plants photosynthesis is among the first processes being affected by elevated temperatures. The damage due to heat stress (HS) includes a wide range of structural and functional changes. The results from in vivo and in vitro studies showed that the primary target of high temperature stress (HTS) involves the PS2 reaction center (PS2 RC) with reduced water oxidising activity being the major injury. When temperature rises above 35°C, the PS2 core complex, as well as part of LHC2 migrate laterally from the integral spaces in the stroma lamellae where PS1 is mainly located. These thylakoid membrane structural rearrangements are completely reversible when the temperature decreases. It is believed that their physiological role is to protect PS2 from overexcitation under high irradiances.

At higher temperature more extensive breakdown of PS2 occurs, leading to a separation of the PS2 core from the main LHC2, a general destacking of the thylakoid membranes and the phase separation of non bilayer forming lipid components. These latter changes are accompanied by a dramatic loss of PS2 electron transfer activity, which in turn, is reflected in major changes in photochemistry and fluorescence yield of chlorophyll *a* associated with PS2. It is believed that thermal denaturation of PS2 is linked to major physical changes occurring in the lipid matrix of thylakoid membranes during heating which are likely to alter lipid protein interactions and hence cause conformational changes in thylakoid proteins.

The surface charge of the thylakoid membrane is the major property determining its stability. If the charge density on the membrane surface increases, the electrostatic repulsion between equally charged molecules grows as well. As a consequence, lipids become more disordered which could lead to protein lateral diffusion in the membrane plane, accompanied by decrease in its stability, making it more sensitive to heat damage. Direct result of heat inactivation is also a release of functional manganese ions from the PS2 complex. The heat stress induced hyperfluidisation of thylakoid membranes which affects lipid protein interaction and causes various perturbations (phase transitions of lipids as well as conformational changes and destructive processes in the thylakoid membrane) (TM) and altering their major function.

The oxygen-evolving complex activity, energy migration, electrochromic shift of pigment absorption at 515nm, electron transport and photophosphorylation are among the first to be affected and injured. The breakdown of $O_2$ evolution has been associated with damage of PS2. Simultaneously the biosynthesis and radiant energy activation of Rubisco decreases. Chloroplast photochemical activity decreases after HS but light can modify the heat stability of the photosynthetic apparatus. In isolated chloroplasts incubated at high temperatures, light can produce an injurious effect on the electron transport chain.

In wheat plants thermal injury of in vivo photosynthesis was strongly aggravated by bright light. But evidence is accumulated that rather than causes injury, light can alleviate the heat induced inhibition of leaf or chloroplast photosynthesis.

In contrast, when the leaves were exposed to HT in the presence of light no inhibition of oxygen evolution and fluorescence was detected, clearly showing the protective nature of light against heat injury of these processes. Based on the chlorophyll fluorescence parameters behaviour at HT stress we try to present in consecutive order the injury to different parts of the photosynthetic machinery. At high temperature conditions the changes in different parameters of chlorophyll fluorescence induction kinetics are caused by some processes proceeding simultaneously, but dominating at different temperature diapasons. They influenced differently $F_v$, $F_0$ and $F_m$.

At mild temperatures 25-40°C the main influence on fluorescence changes is connected with two processes: enhancing the radiationless deactivation of excited chlorophyll molecules and increasing the excitation energy migration from LHC2 RC2 to RC1. The latter leads to an increase in the size of weakly fluorescing PS1. Both processes decreased the quantum efficiency of fluorescence resulting in an amplitude lowering of all three parameters. More complex is the situation in the 40-47.5°C range, where besides the above mentioned processes proceed, in addition, at least three other processes, modifying fluorescence quantum yield directly and changing the rate of electron transport around PS2.

Part of the LHC2 complexes dissociated from the PS2 core and are in a free state so they could realise chlorophyll excitation only through irradiation or radiationless deactivation, which leads to an increase of $F_0$ fluorescence. Simultaneously an inactivation of oxygen evolution system occurs which leads to formation of a fluorescence quenching state (e.g. $P^+_{680}$), and causes the sharp drop of $F_v$ in the 40-45°C range. These temperatures provoke also the changes in the acceptor side of PS2. In this situation a part of PS2α centres is transferred in PS2β $Q_B$-non reducing ones.

At very HT (above 52°C) a denaturation of protein components of LHCP occurs which changes the surrounding of chlorophyll molecules and increases the access of the molecule quenchers to excited chlorophylls. As a result fluorescence is almost entirely inhibited. Fortunately, organisms are able to acquire thermotolerance and to survive within a wider temperature interval involving constitutional and facultative changes. The later are transitional phylogenetic modifications that appear in response to different environmental agents.

## Temperature Acclimation of Plants

### *Protein Synthesis*

Acclimation involves both short-term chemical, molecular and physiological responses and long term physiological structural and morphological modifications. In most cases the adaptive mechanisms to different thermal regimes could be considered as compensatory, because they enable plants to buffer the effect of the temperature shift on their metabolic systems. As a result, the amount of specific components compensating the temperature effect on the rate of a given reaction changes. The protein synthesising system plays a crucial role in plant acclimation processes. It has been hypothesised that particular proteins whose synthesis is induced by stress conditions are critical for the survival in that stress.

Pretreatments which lead to acquisition of thermotolerance are conditions under which heat shock proteins (HSPs) are synthesised. During HS these HSPs could interact with other proteins to prevent their aggregation and facilitate reassemblance of functional structure. These and other correlative data strongly suggest, but still do not demonstrate definitively, that accumulation of HSPs is crucial in protecting from thermal killing. If the growth temperature of plants is increased beyond a particular point, dramatic changes in gene expression and translation occur. Instead of normal pattern a new set of proteins is synthesised from new mRNAs.

Munro and Pelham suggest that it is the level of denatured proteins in the cell which triggers HS response. The authors found support of this suggestion in the behaviour of HS induced HSP70 which is involved in degradation of denatured proteins. According to Pelham during HS, HSP70 binds to the exposed hydrophobic regions of protein aggregates and facilitates their degradation. Moreover, some authors called free HSP70 "cellular thermometer". Since it appeared that thermotolerance, in most cases, correlates closely with HSP synthesis it has been hypothesised that the function of at least one HS gene is implicated in causing thermotolerance.

The requirements for protein synthesis in the development of thermotolerance has been shown by inhibitor studies. Addition of cycloheximide immediately prior HS blocks both HSP synthesis and the induction of thermotolerance. The exogenous cytokinin, benzyladenine, which stimulates the protein synthesising system activity affects resistance positively. Accumulating evidence indicates that through the stabilisation of proteins in a particular state HSP90, HSP70 and HSP60 facilitate a wider diversity of important processes including protein folding, transport of proteins across membranes and their correct conformation, the assembly of oligomeric proteins and modulation of receptor activities, DNA replication and mRNA turnover. All these functions require the alteration or maintenance of specific polypeptide conformation.

Based on these activities, HSP90, HSP70 and HSP60 have been termed "molecular chaperones" or polypeptide chain binding proteins. Therefore, chaperone proteins serve multifunctional roles. May be they prevent the formation of incorrect structures or support reconstitution of partially denatured proteins. The HSP60 in chloroplasts is the Rubisco subunits binding protein also known as chloroplast chaperonin. It is proposed to be involved in assembling the Rubisco holoenzyme. In addition to high molecular weight (HMW) HSPs typical of eucariotes, higher plants and algae also synthesise nucleus encoded low molecular weight (LMW) 17-26kDa proteins which are localised in chloroplasts.

In contrast to HMW HSPs presented both constituively and at elevated temperatures only certain LMW HSPs are present in absence of stress. Some of LMW HSPs have been demonstrated to became associated with photosynthetic membranes, potentially protecting the photosynthetic apparatus from damage at high temperatures. The fact that chloroplast HSPs are found in diverse plant species and that chloroplast HSP mRNA is abundant during stress suggests that those proteins are major components of the HS response.

The half-life of chloroplast HSP21 in pea at control temperatures was 52±12h suggesting that the protein function is critical during recovery, as well as during stress. There is also an evidence for thermal adaptation dependent regulation of HSP21 synthesis. It was shown that in pea, bean, Chlamidomonas and some other plants HSP21, which is the only heat-induced chloroplast protein, is bound to chloroplast membranes. Such a localisation would be in accordance with its probable function against photoinhibition under HS conditions. In agreement with this interpretation HSP22 in Chlamidomonas was found in the grana thylakoid membrane fraction.

How LMW HSPs fulfil their protective role is not known. In the laboratory of Kloppstech it was proposed that the chloroplast HSP21 functions to protect or repair PS2 which is one of the most sensitive chloroplast components during stress. In spite of the great interest in HSPs it is still not possible to say categorically that their and only their synthesis is responsible for the acquisition of thermotolerance. It is also possible that their role is not in the development of thermotolerance, but in the repair of heat induced damages.

### *Role of Lipids*

The possible role of plant lipids, particularly membrane lipids, in high temperature susceptibility and tolerance is also an important problem. The degree of unsaturation of acyl residues of glycerolipids determines the physical characteristics of membranes. Therefore one can be postulated that the degree of fatty acid unsaturation should affect various functions of membrane bound proteins. Glycerolipids of the thylakoid membrane not only serve as a major constituent of the membrane-forming bilayer but they also provide hydrophobic ligands to membraneous proteins. They play an important role also in maintaining the photosynthetic electron transport machinery. As Williams noted the role of lipids in determining the ability of plant to resist thermal stress is not well known. But, a remarkable increase in protein/lipid ratio under acclimation was observed in thylakoids.

High temperature acclimation of higher plants is usually, but not always, accompanied by increased lipid saturation. Many authors observed that the rise in growth temperature increases the level of saturated fatty acids in membrane lipids and enhances the heat stability of photosynthesis. Moreover, Thomas et al. found marked increases in PS2 stability following hydrogenation of the thylakoid membranes of pea chloroplasts. In the same time, Gombos et al. have reported little or no difference in the thermal stability of a fatty acid desaturase mutant of *Synechoccystis* PCC6803 that was incapable of synthesising polyenoic lipids and wild-type cells, suggesting that the presence of such lipids plays little part in PS2 ability in cyanobacteria.

High temperature acclimation involves also adaptive changes in lipid composition. Raison et al. showed that during heat acclimation the threshold temperature at which fluidity still maintains the native membrane structure and function rises. We have also demonstrated that thermotolerance of the photosynthetic apparatus is influenced by the ratio of the main lipid components of thylakoid membrane i. e. MGDG/DGDG. This ratio decreased from 1.3 in control bean plants to 0.9 in heat adapted plants.

Moreover, we tentatively identified DGDG with fatty acyl pairing 18:1/16:0 besides those species formed by control plants at 25°C as being accumulated by heat induction in thylakoid membrane of heat adapted but not in notadapted plants. Therefore, the achieved thermotolerance of the majority of thermosensible light reactions is probably caused by both the lipid induced more stable conformations of membrane connected protein subunits and the adjustment of membrane lipid fluidity.

## Low Temperature Stress

The reaction of photosynthetic apparatus to low temperatures as well as to water and salt stresses shows some similarities to HS response. It is also characterised by alteration in thylakoid membrane electron transfer connected with PS2, the oxygen evolving complex and photophosphorylation. Biomembranes are the main target for attack at low temperature and during freezing. The synthesis of some key proteins required for photosynthesis is specifically susceptible to low temperature stress. The prolonged exposure of leaves to low temperature results in selective inactivation of the oxygen evolving system in cucumber, bean.

It was shown that during cold acclimation newly translatable mRNAs were induced and new set of proteins were synthesised i.e. different gene expression occured. Low temperature and cold acclimation modify PS2 RC and LHCP. According to Krupa et al. the specific changes in lipid composition account for the structural differences between cold hardened and non hardened LHCP2. However, not all plant species have shown a decrease in trans hexadecenoic acid at low temperatures. Acclimation of winter rye to cold temperature results in an increased capacity for photosynthetic electron transport, a decrease of the apparent activation energy of recombination and in an increased resistance to low temperature induced photoinhibition (PI). Cold acclimation caused also changes in the activity and isozyme variations of some enzymes. In cabbage it induced changes in Rubisco, leading to the formation of two distinct forms of two subunits.

### Role of the Xanthophyll Cycle

In conditions of strong light or combined action of intense light and high or low temperature the carotenoid composition of thylakoid membrane responded rapidly with zeaxanthin (Zx) being formed by violaxanthin (Vx) deepoxidation and recovered to Vx in low light. It is believed that this reversible formation of Zx in the so called xanthophyll cycle (XC) protected the chloroplasts against harmful effects of excessive light. As Havaux and Gruszecki mentioned the mechanism by which Zx could exert its supposed protective role remains however illusive. Its exact physiological function is still not clear, but several proposals suggested that it takes place in thermal dissipation of the excess of radiation energy absorbed by the photosynthetic pigments. It is shown that energy of the lowest singlet excited state is higher than the energy of $Q_y$-state of Chl. *a* in the case of Vx but lower in the case of Zx.

A relative position of these levels on energy scale favours a singlet-singlet excitation energy transfer from Vx to Chl. *a* (antenna function) and from Chl. *a* to Zx (chlorophyll excitation photoprotection). The structural effect of Zx with respect to the thylakoid membrane is important in the physiological response to stress conditions. Havaux and Gruszecki observed Zx related increase in thermal resistance of PS2. As can be seen from Fig. 1, at the presence of Zx the thermotolerance of Chl. *a* fluorescence of potato leaves increased considerably. The temperature at which thylakoid membrane was irreversibly injured increased with 3-5°C.

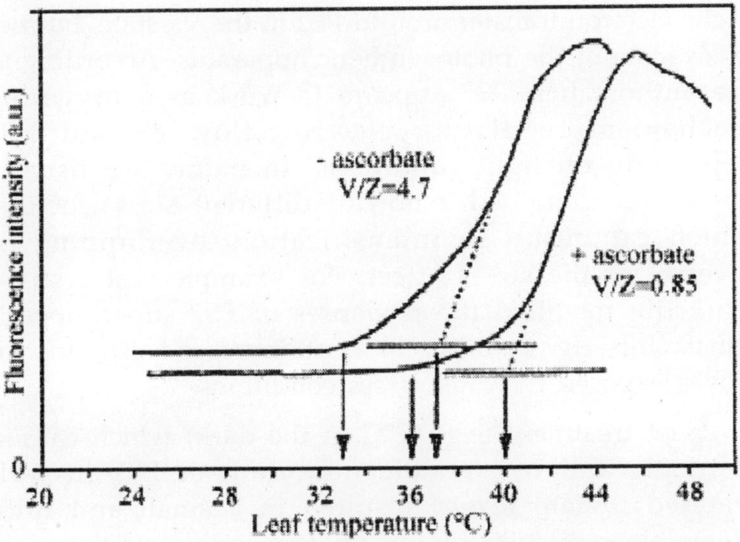


*Figure 1. Temperature dependence of the chlorophill a fluorescence intensity (in low exciting light) in potato leaves containing high or low levels of zeaxanthin obtained by infiltrting them for 2.5 h in the dark with a buffer with or without 100 mM ascorbate, respectively.*

A Zx-induced decrease in lipid fluidity will reduce membrane permeability to small molecules, e.g. oxygen. This effect could be of great importance in respect to the photoprotection of chloroplast membranes as lowered oxygen concentration will decrease destructive reactions involving oxygen, such as lipid peroxidation. Gruszecki et al. postulate also a new physiological function of XC, namely the regulation of the cyclic electron transport around PS2. In their model Vx would play a role in controlling excitation energy flow from accessory xanthophyll pigments towards PS2-RC-β-carotene.

β-carotene was incorporated in this model to the cyclic electron transfer chain and its excitation state was postulated to stop the possibility of such a transfer. Blue light absorbed and transferred by carotenoids may inhibit cyclic electron transfer around PS2 in the Vx state, but not in Zx state of the photosynthetic apparatus. According to the authors here XC appears to work as a triggering mechanism regulating electron flow around PS2 depending on light conditions. In nature we usually observe a combined action of different stress factors, which can result in intensification, overlapping or reversal of the stress effect. For example water stress markedly modified the responses of PS2 superimposed constraints. Its stability to heat increased strongly in leaves exposed to water stress conditions.

Heat treatment (e.g. 42°C in the dark) which caused complete and irreversible inhibition of PS2 in well watered tomato leaves resulted in a small and fully reversible reduction of the photochemical efficiency of PS2 in drought stressed leaves. In this case stress induces a shift of more than +5°C in the resistance of PS2 evaluated on the basis of quantum yield. Increased stability of PS2 to heat by drought stress is of great ecophysiological significance because both stresses are usually combined in the field. Havaux proposed several mechanisms concerning water stress protection to PS2 against heat injury.

Considering the apparent correlation between heat tolerance of PS2 and chloroplast lipids one can propose that water stress increased the stability of PS2 to heat by strengthening the interaction between PS2 proteins and their lipid environment. This may be achieved via the alteration in lipid composition of the thylakoid membrane. Kaiser also reported that high osmotic potential partially prevents photosynthesis from inactivation at supraoptimal temperature and suggested that intracellular salt concentration could be an important agent for adaptation to high temperature.

## Role of Plant Growth Regulators

Plant growth regulators play an important role in plant behaviour, especially in stress conditions. Stress can change the levels of specific hormones or the plants sensitivity to them. As Morgan mentioned, from the days of Darwin and of Went plant hormones have been viewed as chemical messagers regulating the normal progression of developmental changes, as well as responses to environmental signals. Abscisic acid (ABA), proposed as a common mediator of plant stress responses, is associated with cold, salt, drought and heat tolerance. According to Skriver and Mundy plant stress can also be induced by exogenous ABA.

A family of dehydrins accumulate in a wide range of plant species in response to dehydratation stresses and ABA treatment. A novel property of some ABA-responsive proteins and dehydrins is their resistance to heat induced coagulation. It has been hypothesised that the role of these unique proteins is to function in combination with cell osmolytes, such as sucrose, to prevent denaturation and coagulation of cellular proteins and membrane under stress conditions.

Ivanov et al. showed that exogenous ABA induced in barley seedlings protection of PS2 against photoinhibition and this effect was accompanied by higher photochemical quenching. As with many other stresses growing plants at high temperatures increased their ABA content and often conjugated ABA as well. It was shown by Robertson et al. that the bromegrass cell suspension culture in medium containing 75μM ABA without prior heat treatment had a 87% survival rate, as determined by regrowth analysis following exposure to 42°C for 120 min.

In contrast, less than 1% of the control cells survived during this heat treatment. It was also shown that sucrose (8%), in combination with responsive to ABA heat stable proteins, is the most effective in conferring heat stability. As it is known polyamines are an important factor regulating growth, protein biosynthesis as well as stabilising chloroplast thylakoid membranes and retarding chlorophyll degradation. Polyamine metabolism change may play a role in plant adaptation to agents including biological stresses and may serve as a homeostatic buffering mechanism to stabilise cellular pH in stressed plant cells.

*Table 1. Influence of PAs on the Fv/F0 and Fv/Fm ratio of chlorophyll fluorescence in leaf discs and isolated chloroplasts from control (C), acclimated (A) and non-acclimated (NA) plants*

| Variants | Leaf discs | | | | Chloroplasts | | | |
|---|---|---|---|---|---|---|---|---|
| | Fv/F0 | | Fv/Fm | | Fv/F0 | | Fv/Fm | |
| Put | – | + | – | + | – | + | – | + |
| C | 5.69 | 5.17 | 0.85 | 0.84 | 1.77 | 3.19 | 0.64 | 0.76 |
| A | 4.26 | 4.30 | 0.81 | 0.81 | 3.86 | 4.85 | 0.79 | 0.83 |
| NA | 3.44 | 3.32 | 0.78 | 0.77 | 1.75 | 2.85 | 0.64 | 0.74 |
| Spd | – | + | – | + | – | + | – | + |
| C | 4.67 | 4.64 | 0.82 | 0.82 | 1.44 | 3.10 | 0.59 | 0.76 |
| A | 4.63 | 4.78 | 0.82 | 0.83 | 2.78 | 3.68 | 0.73 | 0.78 |
| NA | 3.96 | 4.19 | 0.80 | 0.81 | 1.85 | 2.85 | 0.64 | 0.74 |

Many of the biological functions of polyamins (PAs) appear to be attributed to the cationic nature of these molecules, which are highly protonated at physiological pH. On Tables 1 and 2 the data about the influence of PAs on $F_v/F_0$ and $F_v/F_m$ ratio in the leaf discs and isolated chloroplasts from control, acclimated and non acclimated bean plants are shown. It could be seen that the treatment of leaf discs with putrescine as well as with spermidine did not change considerably the induction kinetics pattern and the amplitude of the prompt chlorophyll fluorescence at 25°C. Polyamines strongly influenced the fluorescence parameters of isolated chloroplasts. That was especially characteritic for the parameter $F_v/F_0$.

*Table 2. Influence of PAs on T50 of Fv/F0 ratio in leaf discs and isolated chloroplasts from control (C), acclimated (A) and non-acclimated (NA) plants*

| Variants | Lcaf discs | | Chloroplasts | | Leaf discs | | Chloroplasts | |
|---|---|---|---|---|---|---|---|---|
| | -Put | +Put | -Put | +Put | -Spd | +Spd | -Spd | +Spd |
| C | 45.5 | 45.8 | 33.3 | 36.3 | 47.7 | 47.0 | 33.3 | 37.3 |
| A | 47.3 | 47.5 | 34.7 | 37.4 | 49.5 | 49.0 | 35.7 | 40.1 |
| NA | 48.3 | 48.0 | 33.7 | 35.8 | 50.0 | 49.0 | 32.9 | 36.5 |

Putrescine increased it with 80% in chloroplasts from control plants. Analogous was the spermidine action increasing the same parameters with 114% in control plants and 54% in non-acclimated ones. The $F_v/F_m$ ratio changed in a similar way. Chloroplasts isolated from the three groups of plants manifested a similar behaviour to temperature inactivation, but for acclimated plants $T_{50}$ was slightly increased. The addition of Put or Spd causes in isolated choroplasts of all three plant groups an increase of $T_{50}$ with 3-5°C.

# 3

# Synthetic Plant Promoters

Plant promoters direct controlled local gene expression in response to pathogens. These promoters could be used to help define signalling pathways, to isolate novel mutants using "targeted genetics", and to engineer plants with increased disease resistance. The control regions of plant genes are modular and contain a number of *cis*-acting elements, each of which may contribute to one or more aspects of a complex expression profile. One strategy to overcome this complexity is to produce synthetic promoters containing only defined individual elements, thereby reducing expression profile complexity.

However, although there are numerous reports of synthetic promoters being inducible by elicitors in transient expression systems, in most cases it is not known to what extent individual cis-acting elements retain their functionality in planta when removed from their native promoter context and whether we can use these individual "modules" to make synthetic promoters that direct a desired expression pattern. Pathogen-inducible promoters represent an attractive system for the production of synthetic promoters. There are a large number of known pathogen-inducible genes, and their promoters are among the best studied in plants. Two groups of pathogen-inducible cis-acting elements, the GCC-like elements and the W boxes, have been well studied.

The GCC box (AGCCGCC) often is found in the promoter regions of defence genes. A similar element has been reported to direct jasmonate and elicitor-responsive expression (JERE; AGACCGCC), and another (DRE; TACCGAC) directs cold-, salt stress-, and dehydration-responsive expression. Recently, another similar GCC-like element called box S (AGCCACC) has been identified that directs expression by fungal elicitors. It appears that minor variations in the core sequences impart responsiveness to different stimuli. The W box [(T)TGAC(C/T)] is the binding site for members of the WRKY family of transcription factors. There is increasing evidence that W boxes are a major class of *cis*-acting elements responsible for the pathogen inducibility of many plant genes.

The importance of W boxes was illustrated recently by studies of the Arabidopsis transcriptome during systemic acquired resistance. In some cases, clustering of W boxes may be associated with inducibility by pathogens. Given that GCC-like boxes and W boxes have been so well studied, it is surprising that there is almost no direct in planta evidence that they can mediate pathogen-inducible expression. Although W boxes have been shown to impart elicitor-inducible expression on a minimal promoter in transient expression systems, there is only one report suggesting that isolated W boxes can function alone in planta and direct pathogen-inducible expression.

It was shown that the W box-containing promoter element E17 mediates gene expression at pathogen infection sites in transgenic Arabidopsis plants. Data concerning GCC-like boxes are equally scarce. A synthetic promoter containing four copies of a GCC box directs ethylene-inducible expression in tobacco, but pathogen inducibility has yet to be shown. Both GCC-like elements and WRKY transcription factors have been implicated in gene expression in response to wounding. It was shown recently that wound- and pathogen-induced signalling consists of networks with some shared components.

It remains an open question, however, whether specific *cis*-acting elements can direct both pathogen- and wound-induced expression in planta or whether these two activities are characteristics of separate elements.

Paul J. Rushton et. al done a study on pathogen inducible synthetic plant promoters constructed from a range of both well-studied and novel *cis*-acting elements. Their study findings are briefly described below. At least seven different elements can alone direct local pathogen-inducible gene expression in transgenic Arabidopsis plants. Major differences are seen between many of the elements with regard to their background expression, their induction by different pathogens, and their speed of induction. Additionally, several pathogen-inducible elements also direct local wound-inducible expression and therefore that components of pathogen and wound-induced signalling are shared.

### Elicitor-Inducible Synthetic Promoters

Our approach to making synthetic promoters that are induced locally by pathogens was first to test candidate elements in a parsley transient expression system for inducibility by a pathogen-derived peptide elicitor, pep25. Promising candidates then were introduced into

Arabidopsis plants to evaluate their in planta expression patterns and inducibility by pathogens. Figure 1A shows how the synthetic promoters were constructed. Each element was inserted between the SpeI and XbaI restriction sites upstream of the – 46 35S minimal promoter of *Cauliflower mosaic virus.*

**A**

| | |
|---|---|
| Box S | C**AGCCACC**AAAGAGGACCCAGAAT |
| GCC box | C**AGCCGCC**AAAGAGGACCCAGAAT |
| JERE | **AGACCGCC**AAAGAGGACCCAGAAT |
| DRE | **TACCGACAT**AAAGAGGACCCAGAAT |
| GCC mut | C**ATCCTCC**AAAGAGGACCCAGAAT |

**B**

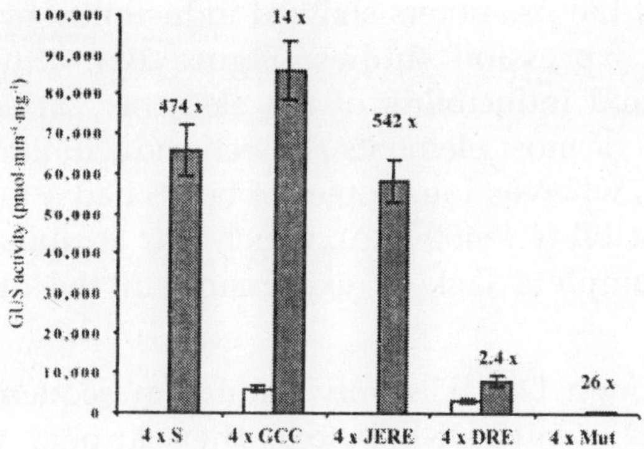


*Figure 1. Elicitor-Inducible Synthetic Plant Promoters. (A) Scheme of the synthetic promoters. (B) Sequence of the elicitor-inducible elements. (C) Elicitor inducibility of the synthetic promoters in a parsley transient expression system.*

Initially, synthetic promoters containing tetramers of candidate elements were constructed. The reason for the use of tetramers was the observation that promoters with multimers of elements are stronger than those with just one or two copies. Because it was unknown whether any of the elements were functional in planta, the researchers used tetramers in initial experiments to ensure that any expression would be strong enough to be detectable by β-glucuronidase (GUS) staining. The first elements tested were boxes W1 and W2 from the parsley PR1 genes, box S from the parsley ELI7 genes, a novel element called box D from the parsley PR2 gene, and an element termed the Gst1 box from the potato gst1 gene.

The Gst1 box contains an S box and a W box separated by just 4 bp in a region of the promoter that mediates transcriptional activation in response to pathogens, during senescence, and in root apices (Figure 1B). All of the promoters showed inducibility by pep25 in transient expression studies (Figure 1C), although the strength and inducibility of the elements varied greatly. Tetramers of most elements showed inducibilities of five- to 30-fold, whereas four copies of box S had a remarkably high inducibility (>400-fold), which was attributable to an almost complete lack of expression in the absence of pep25.

Although box S is very similar in sequence to the GCC, JERE, and DRE boxes, they appear to direct different patterns of gene expression. To investigate these GCC-related elements, they constructed a series of synthetic promoters that are based on box S (Figure 2A). To determine the effect of changes within the core sequence, the flanking sequences remained identical and they altered only the bases necessary to change one variant into another. A construct containing the GCC core sequence was slightly stronger than box S but showed greatly reduced inducibility by pep25 (Figure 2B).

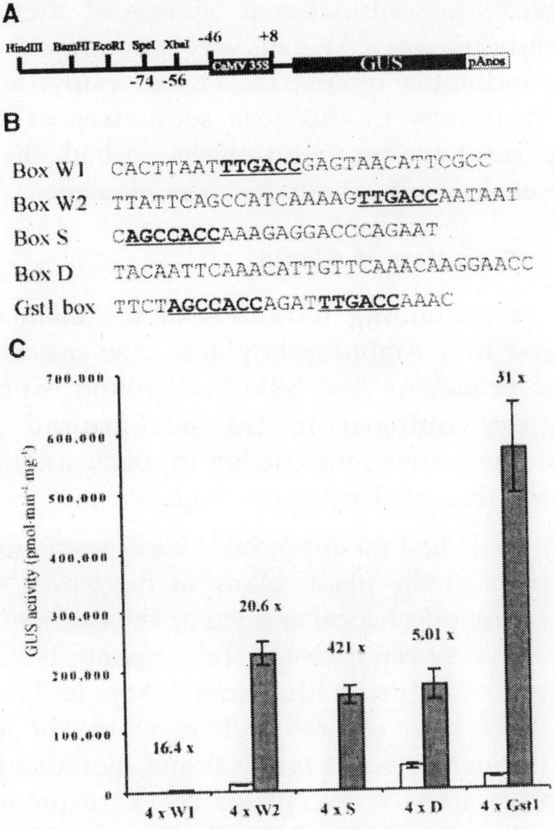


*Figure 2. Elicitor-Inducible Promoters Containing GCC-like Elements. (A) Sequence of the elicitor-inducible GCC-like elements. (B) Elicitor inducibility of synthetic promoters containing GCC-like elements in a parsley transient expression system.*

This reduction to~3% of the box S value is remarkable considering that this is the result of a single base pair change and is attributable primarily to an increase in the background level of the GCC box. Results with JERE were almost identical to those with box S, both in strength and fold induction. In contrast, DRE directed a lower level of expression and showed little elicitor inducibility. A nonfunctional version of the GCC box (GCC mut) (Figure 2A) had very low activity but was slightly inducible by pep25. These results show that minor variations in the core sequences of GCC-like elements can have profound effects on both the strength and the elicitor inducibility of these elements.

### Synthetic Promoters in Planta

Promoters containing tetramers of the elements were introduced into Arabidopsis plants. The majority of the synthetic promoters had little background expression in leaves. By comparison, the background level of expression was sometimes higher in roots, and in the case of 4 × W2, this level was very high.

Only 4 × D had no appreciable background expression in any parts of the plant. Many of the leaves shown in Figure 3A manifest local induction by wounding where the leaves have been excised. This response is very rapid, being induced during the time taken to harvest the plants. Therefore, the researchers wounded leaves by cutting through one-half of a leaf and monitored changes in promoter activity. Except for box D, expression from all promoters was induced locally by cutting.

Figure 3B also illustrates that this massive damage to the leaf sometimes led to a lower level increase in GUS expression over the entire leaf. However, when plants grown in greenhouse conditions exhibited local wound-induced expression, presumably because of insect damage. This finding suggests that local wounding of Arabidopsis plants leads only to highly restricted expression of these promoters.

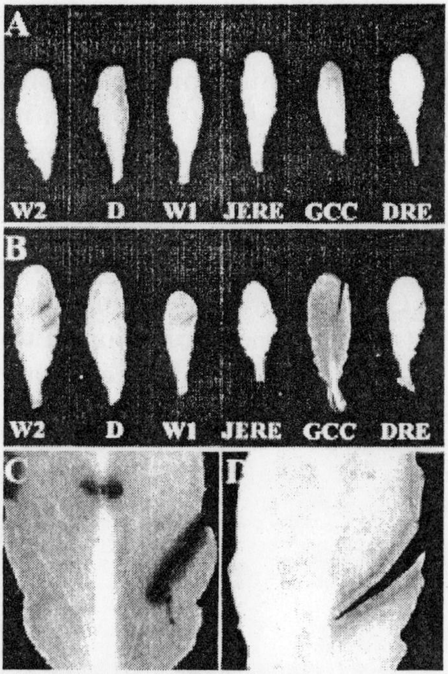


*Figure 3. Expression Patterns of Synthetic Promoters in Planta and in Response to Wounding. (A) Expression patterns (GUS activity) in untreated excised leaves. (B) Expression patterns 1 hr after wounding by cutting. (C) Local wound-induced expression from 4 × GCC 1 hr after wounding. (D) Local wound-induced expression from 4 × S 1 hr after wounding.*

Only when more extensive damage is done is there a response throughout the leaf and subsequent nonlocal induction. The researchers observed two different patterns of local wound-induced expression after cutting. Most common was expression in a layer of cells at the actual cut site. However, with the same promoters, they sometimes also observed local expression in a ring of cells some distance away from the cut site. The exact conditions necessary for the production of each pattern were not investigated further.

They inoculated with the biotrophic oomycete *P. parasitica* pv *Cala2*, which produces an incompatible interaction with Columbia (Col-0) plants. Local gene expression, limited to small areas around infection sites, was seen with 4 × W1, 4 × GCC, 4 × S, 4 × Gst1 (Figure 4), 4 × W2, and JERE. By contrast, no expression was detectable with 4 × D. Figure 4 demonstrates that although 4 × GCC is pathogen inducible, it has a much higher level of background expression in uninfected parts than does 4 × S, and these data accurately reflect the transient expression data described above (Figure 2).

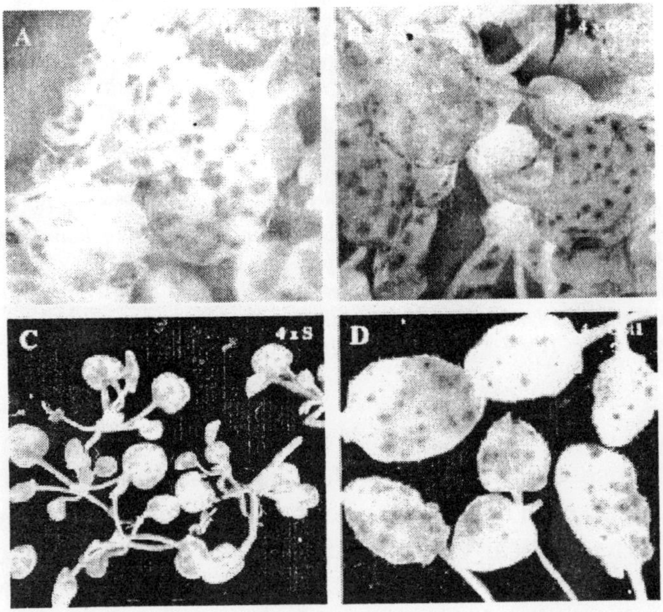


*Figure 4. Local Expression in Response to P. parasitica pv Cala2. (A) Local GUS expression 2 days after P. parasitica infection in rabidopsis plants containing the synthetic promoter 4 × W1. Blue spots represent infection sites. (B) Local expression in 4 × GCC plants. (C) Local expression in 4 × S plants. (D) Local expression in 4 × Gst1 plants.*

For all of the tested elements except box D, these results demonstrate that these elements alone can direct pathogen-inducible expression in planta when removed from their native promoter context.

#### *Local Expression during Powdery Mildew Challenge*

The researchers next investigated the inducibility of the promoters during the nonhost interaction with the barley powdery mildew *Blumeria graminis* f. *sp. hordei*. Only four synthetic promoters (4 ×S, 2 ×W2/2 ×S/2 ×D, 4 ×GST, and 4 ×W2/4 ×S) directed significant local gene expression under the conditions tested. These promoters all contain box S, suggesting that box S may play a role during nonhost signalling. The germinating spores (sp) form appressorial germ tubes (agt), and although most do not penetrate the plant, local gene expression is found predominantly in the underlying mesophyll cells just beneath the attempted penetration site, suggesting that recognition events on the leaf surface trigger defence responses in the underlying cells.

Importantly, the level of expression from these promoters and the number of expressing cells correlated directly with the extent of differentiation of the fungus. This is illustrated in Figure 5B, where at one infection site, haustorium initial formation has triggered cell death. Here, a larger number of cells show activation of the promoter, and levels of expression are higher than in the two other infection attempts, in which infection progressed only as far as papilla formation. During the compatible interaction with the powdery mildew *Erysiphe cichoracearum*, all of the promoters except DRE directed local expression.

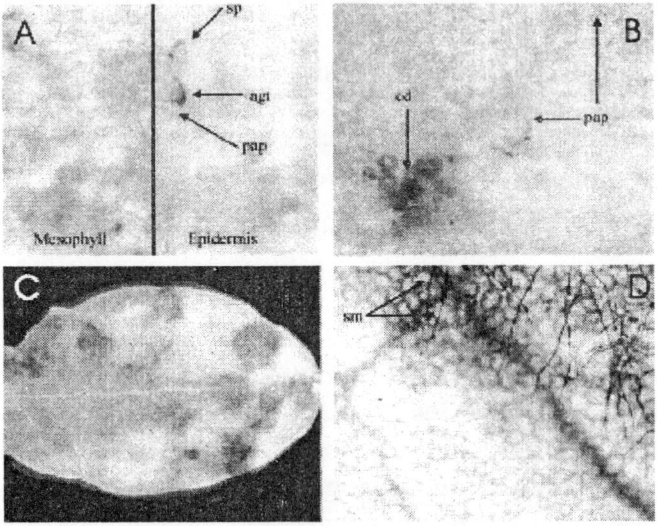


*Figure 5. Local Expression during Nonhost and Compatible Powdery Mildew Interactions. (A) Light micrograph of one leaf infection site on a 4 × S plant viewed at two different planes of focus. (B) Local expression in a 2 × W2/2 × S/2 × D plant upon B. graminis challenge. (C) Local expression 7 days after infection with the compatible powdery mildew E. cichoracearum in a 4 × S plant. (D) Closeup of the border region of an infection site from (C).*

After 7 to 10 days, areas of powdery mildew growth could be seen, and expression of the synthetic promoters was very high in these infected areas. When viewed closely, intense GUS staining was apparent in these areas, and a wave of gene expression was apparent in advance of the growing hyphae.

## Local Gene Expression

They also performed infections with the necrotrophic bacterial pathogen *P. syringae* pv *tomato* using two different isolates: a Col-0-compatible isolate (DC3000) and a Col-0-incompatible isolate (DC3000 carrying the avrRpm1 gene). Figure 6 shows the expression patterns obtained with some of the synthetic promoters. Expression of 4 × W2 was induced locally within 6 hr of infection with the incompatible isolate (Figure 6A). Expression also was apparent with the compatible isolate, although the level of expression was lower. Similar results were obtained with 4 × JERE (Figure 6B), 4 × W1, 4 × S, and 4 × Gst1. Although untreated leaves from plants containing 4 × GCC showed some background expression, often this was increased greatly by mock infection with $MgCl_2$ (Figure 6D); it was difficult, therefore, to determine whether induction by *P. syringae* was occurring.

By contrast, 4 × GCC showed clear induction by *P. parasitica* (Figure 4B), even though background expression was apparent. This finding suggests that the process of injecting liquid into the leaf is itself a sufficient abiotic stress to induce high-level expression of 4 × GCC. Closer inspection of plants containing other synthetic promoters (most notably, 4 × W2, 4 × Gst1, and 4 × W1) showed that background expression levels also are increased sometimes after infiltration of $MgCl_2$. This is similar to the results obtained when a cut was made through one-half of a leaf (Figure 3) and suggests that wounding/stressing a large portion of a leaf can lead to a response in the entire leaf.

Infection of 4 × D-containing plants led to qualitatively different results. No expression was detected in control or mock-infected leaves or with either isolate 6 hr after infection. By contrast, strong local gene expression was seen 24 hr after infection with both isolates. Again, the level of expression was higher during the incompatible interaction.

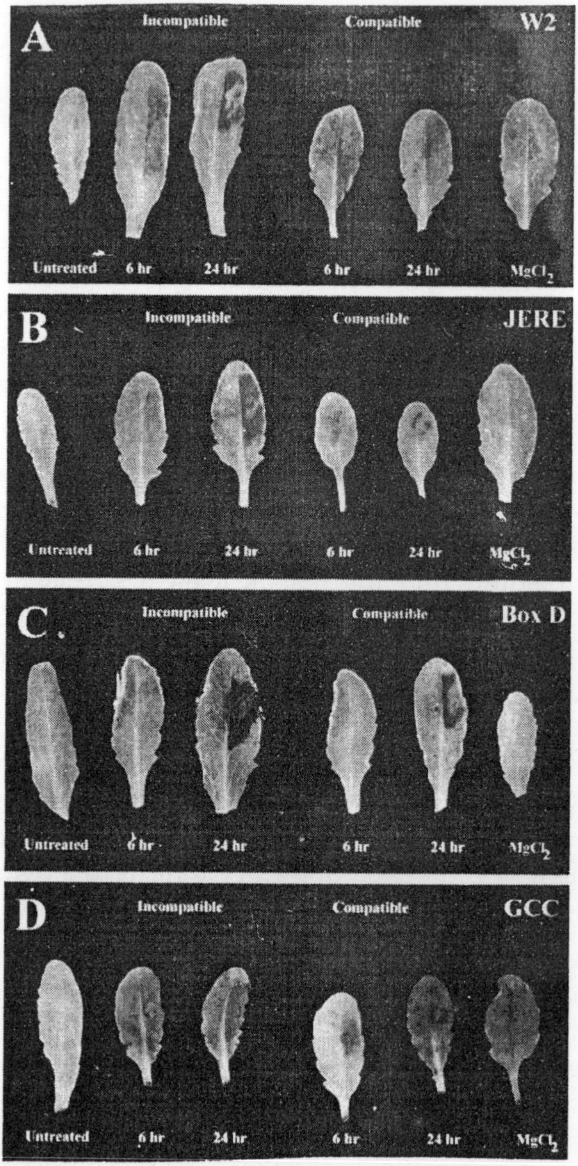


*Figure 6. Expression during Interactions with P. syringae. (A) 4 × W2. (B) 4 × JERE. (C) 4 × D. (D) 4 × GCC.*

**Element Number on Strength and Inducibility**

Although the synthetic promoters containing tetramers of elements direct local gene expression upon pathogen attack, many are not yet ideal for our purposes because they have more background expression than desired or they respond to a number of biotic and abiotic stresses. Therefore, they started to construct improved "second-generation" synthetic promoters using these *cis*- acting elements as building blocks.

The first parameter investigated was the effect of the number of copies of an individual *cis*-acting element in a synthetic promoter. The researchers constructed a series of promoters containing one, two, four, and eight copies of an element and tested these by transient assay. Figure 7A shows the results obtained with 1 × W2, 2 × W2, 4 × W2, and 8 × W2. Increasing the number of copies of W2 increased the strength of the promoter progressively. However, 2 × W2 had the best inducibility, because additional copies caused a proportionally greater increase in background expression.

Similar results were obtained with boxes S and D. These findings suggest that promoters containing fewer copies of an element may be better suited to mediate pathogen-specific inducibility in planta. Accordingly, they introduced promoters with two copies of elements into Arabidopsis plants. Figures 7B and 7C show a comparison of 2 × S and 4 × S plants treated with *P. syringae* and demonstrate that the differences between the two promoters that were observed in the transient assay also are apparent in planta. Both promoters were inducible by pathogens, but 4 × S showed a higher background level and clear wound induction where the leaf was excised. By contrast, 2 × S, although clearly

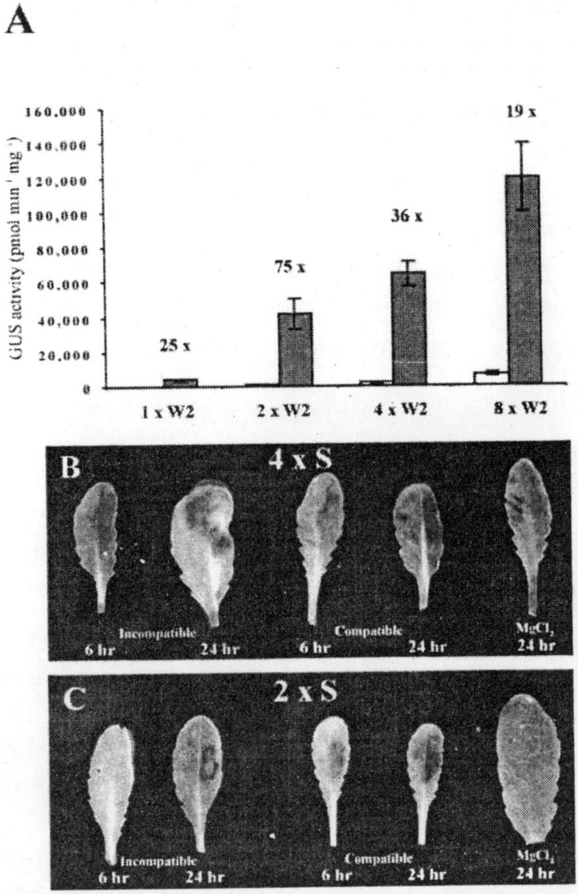


*Figure 7. Effect of Element Number on Strength and Inducibility. (A) Elicitor inducibility of synthetic promoters containing increasing numbers of the elicitor-inducible cis-acting element box W2. (B) and (C) Expression patterns with plants containing the synthetic promoters 4 × S (B) and 2 × S (C) during interactions with P. syringae.*

pathogen inducible, showed on average less background and in many cases no apparent wound induction.

### Spacing Effects and Promoter Strength

Our data show that not all W box-containing synthetic promoters behave similarly. For example, both transient expression experiments (Figure 1) and results from transgenic plants show that box W2 is much stronger than box W1, even though both contain the same TTGACC core element. To eliminate possible spacing effects, they made new versions of both box W1 and box W2 that are identical in length and have TTGACC core sequences in identical positions (Figure 8A).

The only difference between the new version of box W1 and the original is the addition of the preceding 7 bp from the PR1-1 promoter. Both of the new elements were strong, with no detectable difference between 4 × W2, 4 × W2 new, and 4 × W1 new (Figure 8B). The seven additional bases in box W1 new have increased the strength either by adding a positive element or by altering the spacing between TTGACC core elements. Therefore, they made a version of 4 × W1 new in which an unrelated sequence was substituted for the original seven bases (Figure 8A). This construct (4 × W1 5 prime mut) directed a similar level of expression to 4 × W1 new (Figure 8B).

Thus, the difference in strength between the two versions of box W1 was not sequence dependent, excluding the possibility that a new cis-acting element was generated. The pronounced difference in strength between 4 × W1 and 4 × W2, therefore, seems to be a spacing effect caused by different distances between core TTGACC elements. To establish whether spacing from the TATA box had any effect on promoter strength or inducibility, they inserted elements 18 bp farther upstream into the SpeI site instead of between the SpeI and XbaI sites (Figure 1A).

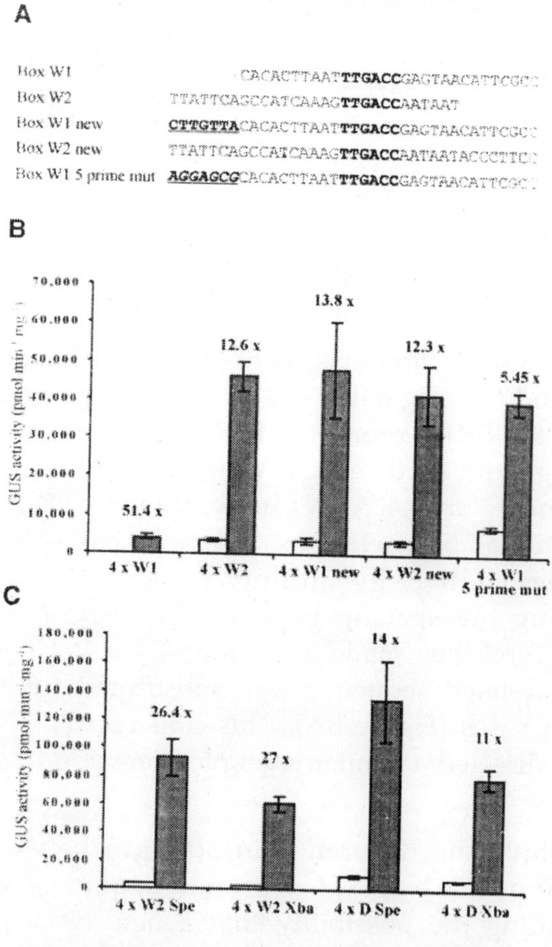


*Figure 8. Spacing Effects and Promoter Strength. (A) Sequences of the different versions of box W1 and box W2. (B) Elicitor inducibility of the synthetic promoters shown in (A). (C) Elicitor inducibility of synthetic promoters with elements inserted at different positions.*

Figure 8C shows that this resulted in a small increase in strength but had little effect on inducibility. Placing the elements at other positions farther upstream had little effect. Together, the results suggest that positioning an element closer to the TATA box can result in some change in promoter strength but that there is little effect on inducibility.

### Box D and the Role of Flanking Sequences

Box D is of considerable interest because box D-containing synthetic promoters have no appreciable background expression or wound inducibility in planta, and they also show different kinetics of induction by pathogens. Box D was discovered as a DNase1 footprint from approximately -76 to -52 in the parsley PR2 promoter. This region (Figure 9A) had a high elicitor inducibility but was weak (Figure 9B). Because the exact extent of the element was unclear, a longer version containing the next six bases from the PR2 promoter at the 3′ end was constructed (Figure 9A). Box D was almost 30 times stronger than 4 × D short (Figure 9B), although inducibility was reduced as a result of increased background levels.

They then constructed a version of box D in which the six additional bases were exchanged, and this was almost identical to box D short (Figure 9A), demonstrating that the considerable increase in strength is a sequence-specific effect that is dependent on the six bases GGAACC.

Therefore, they believe that box D consists of at least two elements: an elicitor-responsive element and a positive-acting element. This positive element cannot function alone but appears to be a coupling element that forms a functional unit with the elicitor-inducible element(s). Alternatively, the shortened version of box D simply may lack crucial 3′ nucleotides of a single element required for selective high-affinity binding of a specific transcription factor, thus allowing binding to other related factors, leading to the observed differences in inducibility.

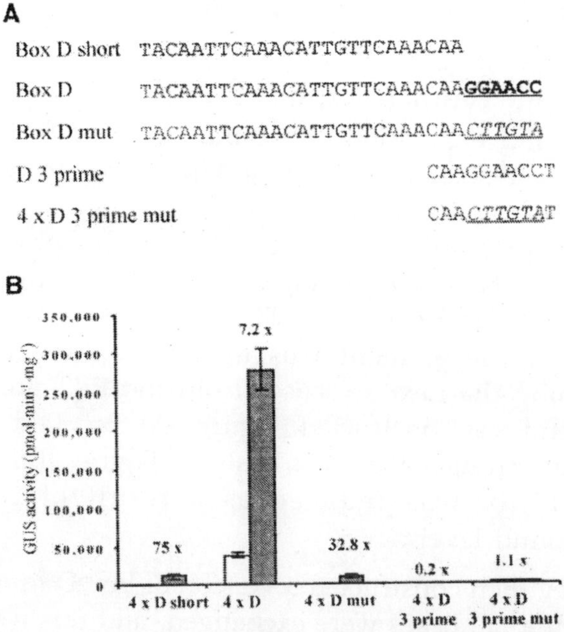


*Figure 9. Box D. (A) Sequences of the different versions of box D. (B) Elicitor inducibility of the synthetic promoters shown in (A).*

Having taken a reductionist approach, reducing synthetic promoters to a single type of cis-acting element to demonstrate functionality, they next took the first steps toward making improved synthetic promoters that contain more than one type of defined cis-acting element. In contrast to the transient expression results, in planta studies produced encouraging results. They introduced two synthetic promoters (2 × W2/2 × S/2 × D and 4 × W2/4 × S) that contain combinations of different elements into Arabidopsis plants. Both showed good inducibility by a range of pathogens and were among the best promoters tested, combining high inducibility with low background. It seems likely that all of the cis-acting elements contribute to the overall expression of the synthetic promoter and that promoters containing carefully selected combinations of elements may be among the best pathogen-inducible promoters.

### Pathogen-Inducible Synthetic Promoters

A range of pathogen-inducible cis-acting elements can alone mediate pathogen-inducible expression in planta. When taken out of their native promoter contexts, they retain pathogen inducibility as components of synthetic promoters, directing expression that is local and that correlates with the extent of growth of the pathogen. The cis-acting elements tested fall into three groups: W boxes, GCC-like boxes, and box D. The observations suggest that binding sites for WRKY (W box) or AP2/ERF (GCC-like box) transcription factors can be sufficient to confer pathogen inducibility on a promoter. This is an important observation because these represent two of the three largest families of plant-specific transcription factors factors.

It will be interesting to extend these studies to include other pathogen-inducible elements such as Myb Recognition Elements (MREs) and as-1-like elements, although evidence suggests that these elements might not be able to function alone. Interestingly, although more than one type of cis-acting element is not required for pathogen inducibility, some pathogen-inducible promoters contain elements of more than one type. An example is the Gstl box, which contains both a W box and an S box. This places the gstl gene under the control of both WRKY and AP2/ERF transcription factors.

The potato *gst1* promoter has been well studied, and the work provides the first clear evidence of how this gene is activated transcriptionally in response to pathogens. It may be common that signalling pathways operating via different transcription factors can target the same gene; another example is the parsley *WRKY1* gene. The results suggest that defence signalling is largely conserved across species boundaries, because an element from a potato gene is active in a parsley transient expression system and in Arabidopsis plants. In fact, none of the elements tested originate from Arabidopsis promoters, yet all retain their functionality. These results also validate the use of a parsley transient expression system for the initial characterisation of elements, because they successfully identified and characterised numerous pathogen-inducible elements and synthetic promoters in a way that would not have been possible in planta. In almost all cases, elicitor inducibility in the transient expression system is accompanied by pathogen inducibility in planta.

## Signalling in Plant-Pathogen Interactions

Table 1 summarises the responses of the synthetic promoters to pathogens under the conditions tested. Importantly, not all of the elements respond in the same way to pathogens and wounding. Even elements that are bound by the same family of transcription factors (e.g., GCC, JERE, S, and DRE) show different expression patterns. This is consistent with reports that these elements may respond to different hormones, for

*Table 1. Expression Patterns of Synthetic Promoters*

| Synthetic Promoter | P. parasitica Incompatible | P. syringae Incompatible | P. syringae Compatible | E. cichoracearum Compatible | B. graminis Nonhost | Wounding Abiotic |
|---|---|---|---|---|---|---|
| 4 × W2 | + | + | + | + | – | + |
| 4 × W1 | + | + | + | + | – | + |
| 4 × D | – | + | + | + | – | – |
| 4 × GCC | + | ± | ± | + | – | + |
| 4 × S | + | + | + | + | – | + |
| 4 × JERE | + | + | + | + | – | + |
| 4 × GST | + | + | + | + | – | + |
| 4 × DRE | ND | – | – | – | – | + |
| 4 × W2/4 × S | + | + | + | + | – | + |
| 4 × W2/2 × S/2 × D | + | + | + | + | – | + |

(+), high-level induction; (±), lower level of induction attributable to lower induced expression or high background; (-), no expression; ND, not determined.

example, salicylic acid (W boxes), jasmonate (JERE), and ethylene (GCC). Therefore, they generated a spectrum of synthetic promoters useful for studying plant-pathogen interactions at the molecular level.

During compatible interactions, the synthetic promoters accurately report the in vivo situation in which gene activation is slower and perhaps weaker. Under the conditions tested, only promoters containing box S consistently showed local expression during the nonhost interaction with barley powdery mildew. Box S, therefore, could be a useful tool for studying nonhost responses and for engineering plants with broad-spectrum resistance. These results are remarkable considering that there is only a single basepair difference between box S and the GCC box (Figure 2). This single alteration results in box S having much lower background levels both in the transient expression system and in planta. Given that both elements probably are bound by AP2/ERF transcription factors, experiments designed to identify family members with high affinities for box S could yield useful data concerning signalling during plant defence.

Box D is an extremely interesting novel element, unlike any of the others tested, because it combines an apparent lack of background expression or induction by wounding with strong induction during some, but not all, plant-pathogen interactions (Table 1). Box D has been found to be responsive responsive to P. syringae, E. cichoracearum, the bacterially derived peptide elicitor flg22 (data not shown), and the oomycete-derived peptide elicitor pep25 in protoplasts. Box D also responds with different kinetics than other elements, being induced later than the other elements (Figure 6). The molecular characterisation of box D and the identification of its cognate transcription factors would provide new insights into defence gene activation during plant-pathogen interactions.

### Pathogen and Wound Signalling

Recently, work on plant defence signalling has demonstrated the convergence of resistance gene, elicitor, wound, and salicylate responses at the level of mitogen-activated protein kinase (MAPK) activation. It is not clear at present whether these signalling pathways merge at the MAPKs or upstream thereof or whether the same MAPK can mediate disparate responses by interacting with other proteins. The results showing that tetramers of a range of pathogen-inducible cis-acting elements also direct local wound-induced gene expression demonstrate the convergence of resistance gene, nonhost, and wound responses at the level of promoter elements. This observation is in agreement with recent data demonstrating an extensive overlap in the transcriptional response of plants to race-specific elicitors and mechanical stress.

Studies of 290 Avr9/Cf-9 rapidly elicited genes demonstrated that many of these also are induced by wounding. The results presented here extend these observations from the level of the entire promoter to that of individual cis-acting elements. In addition, the identities of the elements suggest that pathways involving salicylic acid, jasmonate, and ethylene may converge at this level. One key question that remains is whether the same or different members of transcription factor families are responsible for both pathogen and wound responses.

In other words, whether convergence occurs at the level of transcription factors as well as at the level of cis-acting elements and promoter architecture. Most likely, these signalling networks have shared components, with proteins at many nodes being capable of receiving inputs from multiple pathways. Many pathogen- and wound-inducible cis-acting elements could represent such nodes.

### Synthetic Promoters

They used synthetic promoters containing tetramers of elements to ensure that the promoters were strong enough to detect activity by GUS staining. These promoters, however, are not optimal for all purposes; therefore, they set about making improved second-generation promoters by varying several parameters. Most important was the number of copies of an individual element in a promoter. Both the strength and the inducibility of a promoter can be modulated by varying the number of copies of an element.

Importantly, this also can have the effect of reducing/ eliminating some background expression because pathogen inducibility appears stronger than basal or wound-induced expression (Figure 7B). Spacing between individual cis-acting elements and/or between these elements and the preinitiation complex also can have a profound effect (Figure 8), but spacing is difficult to predict, and the optimal spacing, like the optimal number of elements, needs to be determined experimentally.

Promoters with combinations of different elements may be among the best pathogen-inducible promoters (Figures 10B and 10C), because they often combine good inducibility with low background. Together, the results suggest that the optimal pathogen-inducible synthetic promoters may consist of combinations of one or two copies of defined cis-acting elements. Perhaps the best synthetic promoters, although made up of carefully

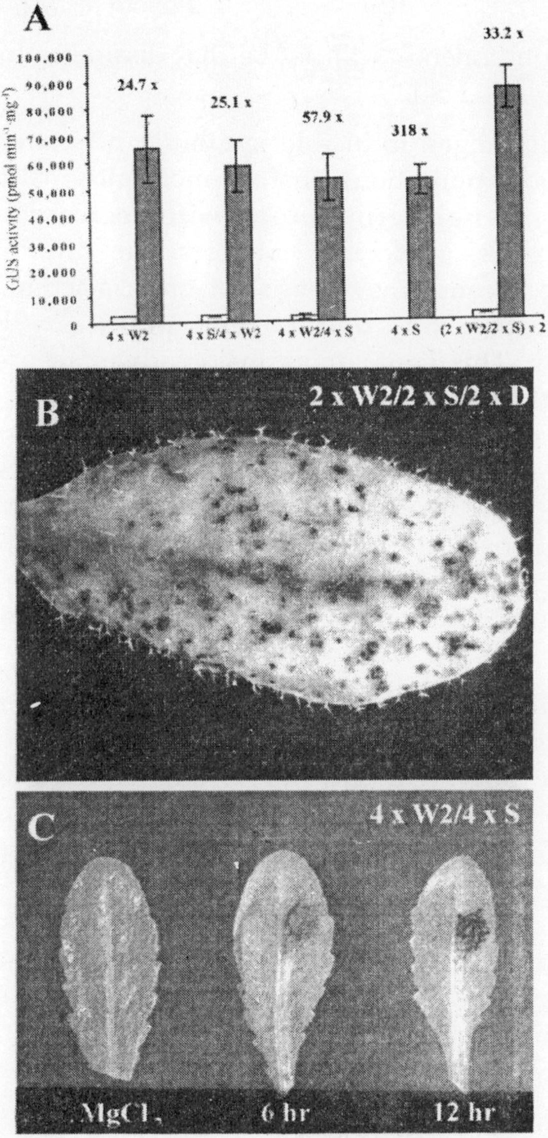


*Figure 10. Synthetic Promoters Containing Combinations of Elements. (A) Elicitor inducibility of synthetic promoters containing combinations of box W2 and box S. (B) GUS expression pattern of the synthetic promoter 2 × W2/2 × S/2 × D 3 days after treatment with B. graminis (nonhost interaction). (C) GUS expression pattern of the synthetic promoter 4 × W2/4 × S 6 and 12 hr after treatment with P. syringae carrying the avrRpm1 gene (incompatible interaction).*

chosen components, will not be that dissimilar to natural promoters after all.

The pathogen-inducible synthetic promoters could have major applications, first, as molecular markers, and second, in engineering crops with increased disease resistance. As molecular markers, the promoters are attractive because they consist of one cis-acting element and therefore one defined end point of signalling pathways. This could have major advantages over the full-length promoters that are used commonly. Through introduction into mutant backgrounds, the synthetic promoters can be used to better characterise these mutants, and the use of defined synthetic promoters for this purpose could become a standard practice. Differences in responses during different plant-pathogen interactions also can be investigated at the molecular level.

Synthetic promoters can be used as reporters for mutant screens using targeted genetics, and this may prove a powerful approach. The spread of plant pathogens and insect pests is increasing worldwide. Researchers have identified numerous plant and pathogen genes that can be used to increase crop resistance toward invading pathogens. These strategies involve interfering with the replication of viruses in the plant, expression of gene products toxic to certain pathogens, and enhancement of the plant's own natural resistance mechanisms.

Such introduced genes usually are placed under the control of strong promoters, yielding constitutive expression of the gene product in all tissues of the plant. This can have detrimental effects on plant growth, development, and crop yield. Use of the synthetic promoters presented here and future improvements thereof may prove valuable in engineering plants with increased resistance, because the use of such defined regulatory sequences may allow highly restricted expression of the desired gene product exclusively at the sites of attempted pathogen invasion. Thus, expression of the gene product is limited to cells surrounding an infection site and is not found in healthy parts of the plant. Moreover, expression via such promoters can be triggered by a range of different pathogens, including during compatible interactions with pathogens of commercial importance, such as powdery mildew. This expression may be sufficient to abort the progression of the invader even in such compatible interactions.

### *Construction of Synthetic Promoters*

Promoter constructs were produced by annealing phosphorylated upper and lower strand oligonucleotides to create elements containing a SpeI restriction site at the 5′ end and an XbaI restriction site at the 3′ end. These were introduced into pBT10-β-glucuronidase (GUS) between the SpeI and *XbaI* sites (Figure 1). Promoters containing multiple copies of elements or combinations of elements in any desired order were obtained by digesting the constructs with either SpeI or XbaI together with SacI, which cuts the plasmid at a site outside of the synthetic promoter. Ligation of two such fragments recreates the plasmid with an increased number of elements. This can be repeated as the 5′ SpeI and the 3′ XbaI sites are recreated, but internal SpeI-XbaI ligations result in the loss of these restriction sites. For analysis in Arabidopsis thaliana, the entire synthetic promoter was excised as a HindIII-SacI fragment and ligated into the binary vector pGPTVGUS-KAN.

### Transient Expression Analysis

Transient expression analysis was performed as described previously using 20 μg of ScaI-linearised DNA per assay. Protoplasts were harvested 8 hr after transfection. All results represent a minimum of seven independent experiments. Normalisation control experiments were performed after cotransfection with a constitutively expressed Petroselinum crispum *UBI4/2*::luciferase construct.

### Transgenic Arabidopsis Lines

*Agrobacterium tumefaciens*-mediated plant transformation was performed as described. Approximately 10 independent lines were isolated for each synthetic promoter-GUS reporter transgene, and between two and four representative lines were subjected to detailed analysis. Tested lines varied between T2 and T5, depending on the promoter tested.

### Wounding of Plants

Leaves were cut with scissors and harvested 1 hr thereafter. Control leaves were excised and immediately stained for GUS activity.

Plants were infected with *Pseudomonas syringae* pv *tomato* strain DC3000 without an avr gene (compatible) or with the *avrRpm1* gene (incompatible) or with *Peronospora parasitica* pv *Cala2* as described and stained for GUS activity. Four- to six-week-old Arabidopsis plants were infected with *Erysiphe cichoracearum* UCSC1 according to the method of Vogel and Somerville and with Blumeria graminis f. sp. *hordei* K1 according to Peterhänsel et al. Clearing of leaves and staining of fungal structures were performed according to Peterhänsel et al.

# 4

# Plant Growth Regulators on Turfgrass Growth

Plant growth regulators has an important role in turf management. Its proper application may improve the quality of turf. Natural products, seaweed extract (SE) and humic acid (HA) are common sources of PGRs that exhibit multiple functions. When applied to turfgrasses, SE enhances root growth, delays senescence, improves turf quality, increases tolerance to drought, and regulates cell membrane components under drought stress. Humic acid, which has auxin-like activity, not only enhances plant growth and nutrient uptake but also improves stress resistance. However, physiological mechanisms of the PGRs' effects on drought tolerance are still not clear.

Plant antioxidants and their role in the plant defence system have received increasing attention within the last decade. Many environmental stresses such as drought, salinity, low temperature, and herbicide pollution damage plants directly or indirectly through the endogenous reactive oxygen species. Water stress causes various physiological and biological changes in plants, one of which is the accumulation of reactive oxygen species in the cell.

The reactive oxygen radicals are toxic and may result in a series of injuries to plant metabolism. It damages photosynthetic components, inactivates protein and enzymes, destroys cell membrane structure and permeability by causing lipid peroxidation. Plants possess various antioxidants to cope with the oxidative stress. These include lipid soluble antioxidants (e.g. α-tocopherol, β-carotene), water soluble reductants (e.g. ascorbic acid, glutathione), and enzymatic antioxidants (e.g. superoxide dismutase, catalase, and enzymes of the ascorbate/glutathione cycle).

Alpha-tocopherol, ascorbic acid, β-carotene, and superoxide dismutase (SOD) are important antioxidants which protect plants by suppressing oxidative injury. Alphatocopherol, which is located within biological membranes, can quench oxygen radicals, stabilise cell membranes by influencing lipid organisation, and protect chlorophyll. Ascorbic acid scavenges many types of free radicals affecting many enzymes activities and also is required for regeneration of α-tocopherol. β-carotene is not only an accessory pigment in the photosynthetic apparatus but also quenches singlet oxygen efficiently. Superoxide dismutase is the most efficient scavenger of the superoxide anion and an essential component of the ascorbate-glutathione cycle for the detoxification of toxic oxygen species.

## Water Stress and Plant Growth

Water is imperative for plant growth and development. Water deficit stress, permanent or temporary, limits the growth and distribution of natural vegetation and the performance of cultivated plants more than any other environmental factor. Although research and practices aimed at improving water stress resistance and water use efficiency have been carried out for many years, the mechanism involved is still not clear. Further understanding and manipulating plant water relations and water stress tolerance can significantly improve plant productivity, turfgrass management, and environmental quality. Water deficit stress can be defined as situations in which plant water potential and turgor are reduced enough to interfere with normal functions.

Smirnoff separates water deficit from desiccation. Water stress is considered to be a moderate loss of water which leads to stomatal closure and limitation of gas exchange. Desiccation is a much more extensive loss of water which can potentially lead to gross disruption of metabolism and cell structure and eventually to the cessation of enzyme-catalysed reactions. Water deficit stress occurs when transpiration exceeds the absorption of water from soil. Deficits may happen under high irrigation when soil water is available because of higher transpiration rate or poor water quality.

As soil water potential declines, daytime wilting occurs when the movement of water toward the roots in the drying soil becomes too slow to replace daytime water loss. Water stress is characterised by reduction of water content, turgor, total water potential, wilting, closure of stomata, and decrease in cell enlargement and growth. Severe water stress may result in arrest of photosynthesis, disturbance of metabolism, and finally dying. Plant growth is defined as irreversible increase of size and is accomplished through cell devision, cell enlargement, and differentiation. It involves genetic, physiological, ecological, and morphological events and their complex interactions. Since plants are rooted in one place, they encounter frequent environmental stresses including water stress.

Water stress influences plant growth at various levels from cell to community. The quantity and quality of plant growth depend on cell devision, enlargement, and differentiation, and all of these events are affected by water stress. Hsiao concluded that water stress inhibits cell enlargement more than cell devision. It reduces plant growth through inhibition of various physiological and biochemical processes, such as photosynthesis, respiration, translocation, ion uptake, carbohydrates, nutrient metabolism, and hormones. As a result, plant size, leaf area, and productivity are reduced. Initially, reduced water potential decreases the chemical activity of water and thereby modifies the structure of water in the cell.

A low chemical activity of water can cause a change in the structure of the sphere of hydration around proteins and thereby reduce their efficacy. A loss of turgor may cause a change in the spatial position of transport channels, membrane enzymes, and decrease membrane thickness. Cell growth (expansion) is one of the most drought sensitive physiological processes due to the reduction of turgor pressure. Cell expansion can only occur when turgor pressure is greater than the cell wall yield threshold.

Water stress greatly suppresses cell expansion and plant growth due to the low turgor pressure. Water stress reduces the rate of photosynthesis. Lawlor pointed out that water stress reduces photosynthetic rate through limiting $CO_2$ diffusion due to the reduced stomatal conductivity, which reduces intercellular $CO_2$ concentration. Mckersie et al. indicated that water stress influences the plants by disruption of membrane structure and concomitant organelle disarray, impairment of stomatal function, and disturbance of $CO_2$ diffusion.

The limitations of carbon dioxide exchange and metabolic fixation result in exposure of chloroplasts to excess excitation energy. Much of this can be dissipated by various photoprotective mechanisms, including dissipation as heat via carotenoids, photorespiration, crassulacean acid metabolism (CAM) idling, leaf movements and other morphological features which minimise light absorption. However, some of excitation energy may be diverted to activate molecular oxygen, and excess reactive oxygen species, will be produced and accumulated in the cell. All these species (e.g. superoxide, hydrogen peroxide, hydroxyl radicals, and singlet oxygen) are reactive and may damage cells by causing lipid peroxidation and inactivation of enzymes.

Plant growth regulators are substances that influence physiological processes of plants at very low concentrations. When produced endogenously by plants, they are often referred to as phytohormones (plant hormones). Plant hormones have been viewed as chemical messengers regulating the normal progression of developmental changes as well as responses to environmental signals. The term PGR includes a large number of synthetic and naturally occurring compounds. Nickell defined PGRs as either naturally or synthetic compounds that are applied directly to a target plant to alter its life processes or its structure to improve quality, increase yields, or facilitate harvesting.

Plant growth regulators play an accepted role in some turfgrass management system. Traditionally, PGRs are used as growth retardents to reduce mowing frequency. Most of these PGRs are synthesised chemicals, and some of them are derived from herbicides. Excessive use of these PGRs has a risk of polluting environments. In recent years, some natural products have received increasing attentions. These naturally derived, and hormone-containing products, can not only stimulate plant growth and development, but also improve plant resistance to environmental stresses. Additionally, these organic substances have no harmful threat to the environment quality. These products, a group of plant growth regulators, are also called biostimulants and include seaweed extracts, humic acids, and vitamins.

### Seaweed Extracts and Humic Acid

Seaweed extracts Seaweed extracts (SE) and seaweed have been used as fertilisers and soil conditioners for centuries. Seaweed extracts contain not only most of the major and minor nutrients, amino acids, and vitamins B1, B2, C, E, but also cytokinins, auxin, GAs, and ABA-like growth substances. Initially, enhanced plant growth with seaweed application was attributed to its soil conditioning properties. Later, it was showed that increased trace element supply could explain only some of the beneficial effects of seaweed.

Low rates of seaweed extract could also promote plant growth significantly. Thereafter, it was suggested that organic compounds, rather than mineral elements, were responsible for improved growth. In recent years, it has been demonstrated that seaweed products contain phytohormones, and the stimulating effects of seaweed extracts may be attributed to these components, especially cytokinins. Seaweed extracts exhibit various beneficial effects on plant growth and development. They may enhance nutrient uptake, regulate plant growth substances, increase chlorophyll content, protein synthesis and cell division, promote root and shoot growth and improve seed gemination.

Seaweed extracts have been used as a cytokinin-like growth regulator and exhibit multiple functions in turf management. When applied to plants, SE stimulates shoot growth and branching, increase root growth and lateral root development, improve nutrient uptake, enhance resistance to diseases and environmental stresses such as drought and salinity. It has been observed that cytokinin or cytokinin-like growth regulators can inhibit the activity of free radical groups, hydrogen peroxide, and superoxide, which are the major elements for chlorophyll degradation during senescence. Under drought stress, they may replace endogenous cytokinin whose production is reduced by stress conditions. It has been indicated that a range of naturally occurring and synthetic cytokinins can enhance stomatal aperture, transpiration, and the protein (polysomes/ribosomes) associated with activity of guard cells.

Humic acid is one of the major components of humus. Humates are natural organic substances, high in humic acid and containing most of known trace minerals necessary to the development of plant life. Humates have long been used as a soil conditioner, fertiliser, soil supplement. Humus holds not just micronutrient metal ions, but also the essential macronutrients: nitrate, phosphate, and potassium. The significance of humic acids is not just limited to its function as a reservoir of mineral plant nutrients and regulator of their liberation.

Recent research showed that humic acid can be used as a growth regulator to regulate hormone level, improve plant growth and enhance stress tolerance. Tan et al. indicated that humic acid (HA) was in general beneficial to shoot and root growth of corn plants. Dry matter yield in corn shoots was stimulated by HA, especially when the plants were treated with HA at a concentration of 640 ppm. Fagbenro and Agboola noted that HA was not only beneficial to the growth but also nutrient uptake of teak seedlings.

The uptake of N, P, K, Mg, Ca, Zn, Fe, and Cu by the seedlings was increased with application of HA. Humic acid produced a proliferation of roots at optimum concentration of 0.05%. These results showed that HA can stimulate plant growth and nutrients uptake at relatively low concentration. The stimulating effects of HA may be related to its hormone activity. O'Donnell found that HA from leonardite exhibits auxin-like effects. It greatly enhanced root initiation of Pelargonium hortorum cuttings.

Piccolo et al. indicated that humic extract with the highest acidity and smallest molecular size is the only material showing auxin-like activity. Several hypotheses have been proposed to explain the effect of HA. These include the formation of complex between HA and mineral ions, catalysis of HA to enzymes in plant, influence of HA on respiration and photosynthesis, stimulation of nucleic acid metabolism, and hormone activity of HA. Recent research showed that HA, when applied to turfgrass, can stimulate shoot and root growth, and improve resistance to environmental stress in turfgrass. This observation is consistent with the results of Piccolo, who showed that HA exhibits an auxin-like activity. However, the mechanism of these two PGRs' stimulating effects on plants has not been well established.

### Water Stress and Reactive Oxygen Species

Increasing evidence suggests that many damaging environmental stresses, such as drought, salinity, temperature, herbicides, etc. have their effects, directly or indirectly, through the formation of reactive oxygen species following impairment of the electron transport system. Water stress is known to cause various physiological and biological effects on plants. The decrease in photosynthesis, the closure of stomata, and osmotic adjustment appear to be typical plant responses to water deficit at the first stage.

Closure of stomata and reduction of photosynthesis in plant leaves disturb the supply of $CO_2$ for photosynthesis and expose chloroplasts to excess excitation energy. When water-stressed leaves were illuminated, the reducing power from the photosystem was not consumed for $CO_2$ fixation but for the activation of $O_2$. Excess reactive oxygen species are accumulated in the cell under drought. Reactive oxygen species are produced by excess transfer energy from triplet excited chlorophyll to oxygen (singlet oxygen formation) or photoreduction of oxygen (formation of superoxide, hydrogen peroxide, and hydroxyl radical).

Excess accumulation of reactive oxygen species results in a series of oxidative injuries to plants. Oxidative injury involves the initial formation of reactive oxygen species and its subsequent reaction with macromolecules. Proteins, lipids, polysaccharides and nucleic acids can be damaged. As a result, normal cell metabolism can be seriously disturbed. Plants possess intrinsic antioxidant defence mechanisms for coping with reactive oxygen species. Essentially, antioxidant defensive systems fall into three general classes:

1) the lipid-soluble, membrane-associated antioxidants (e.g. a-tocopherol, b-carotene);
2) the watersoluble reductants (e.g. glutathione, ascorbate); and

3) enzymatic antioxidants (e.g. superoxide dismutase, catalase and enzymes of the ascorbate/glutathione cycle.

It has been suggested that the formation of reactive oxygen species is an inherent consequence of metabolism and that control of their levels is essential for normal function. The toxicity of an externally imposed biotic or abiotic oxidative stress can be partly attributed to the overriding of existing resistance mechanisms. Only when those mechanisms are overwhelmed would injury occur. This suggests that the strengthening of the defence mechanisms, through enhancing the functions of their components (such as ascobic acid, α-tocopherol, β-carotene and superoxide dismutase) may reduce or prevent oxidative injury and improve water stress resistance of plants.

*Alpha-tocopherol*: Alpha-tocopherol (Vitamin E or 5,7,8 trimethyl-tocol) plays a unique role as an antioxidant and a stabiliser for biological membranes. It is synthesised in chloroplasts and proplastids and located in membranes of the cell, especially the thylakoid membranes of the chloroplasts. The structure and location of α-tocopherol determine its function as a membrane stabliser. Development of HPLC technology facilitates the separation and determination of α-tocopherol accurately. Alpha-tocopherol, concentrated in chloroplasts and present in high concentrations both in chloroplast envelope and in the thylakoid membranes, traps lipid peroxy radicals in the deep inner regions of membranes. It stabilises the bilayer by affecting lipid organisation, and protects chlorophyll against red light-induced degradation. Alphatocopherol is also shown to protect the plant against various environmental stresses such as drought.

*Ascorbic acid (AA)*: Ascorbate (vitamin C) occurs in all plant tissues, usually being higher in photosynthetic cells and meristems (and some fruits). About 30 to 40% of the total ascorbate is in the chloroplast, and stromal concentrations as high as 50 mM have been reported. It is highest in the mature leaf, where the chloroplasts are fully developed and the chlorophyll levels are highest. Although it has been determined that D-glucose is the precursor of L-ascorbic acid, the synthetic pathway has not been totally understood. Ascorbic acid has effects on many physiological processes including the regulation of growth, differentiation and metabolism of plants.

A fundamental role of AA in the plant defence system is to protect metabolic processes against $H_2O_2$ and other toxic derivatives of oxygen. Acting essentially as a reductant and reacting with and scavenging many types of free radicals, AA reacts non-enzymatically with superoxide, hydrogen peroxide, and singlet oxygen. It can react indirectly by regenerating α-tocopherol or in the synthesis of zeaxanthin in the xanthophyll cycle. Therefore, AA influences many enzyme activities, and minimises the damage caused by oxidative process through synergic function with other antioxidants.

*Beta-carotene*: Beta-carotene is present in the chloroplasts of all green plants. It is exclusively bound to the core complexes of PSI and PSII. Protection against the damaging effects of oxygen at this site is essential for chloroplast functioning. β-carotene, as an accessory pigment and an effective antioxidant, plays a unique role in protecting photochemical processes. The carotenoids can exist in a ground state or in one of two excited states after the absorption of light energy. In terms of their antioxidant properties, carotenoids can protect the photosystems in one of four ways:

a) by reacting with lipid peroxidation products to terminate chain reactions;

b) by scavenging singlet oxygen and dissipating the energy as heat;

c) by reacting with triplet or excited chlorophyll molecules to prevent formation of singlet oxygen;

d) by the dissipating the excess excitation energy through the xanthophyll cycle.

The main protective role of β-carotene in photosynthetic tissue may be through direct quenching of triplet chlorophyll, which prevents the generation of singlet oxygen and therefore avoids oxidative stress completely.

*Superoxide dismutase*: Superoxide dismutases (SODs), discovered by McCord and Fridovich in 1969, are ubiquitous metalloenzymes that catalyse the disproportionation of superoxide radical anions to dioxygen and hydrogen peroxide. Superoxide dismutases are the most efficient scavengers of the superoxide anion and an essential component of the ascorbate-glutathione cycle for the detoxification of toxic oxygen species. Higher plants have SODs containing Cu and Zn, Fe, or Mn as prosthetic metals.

Copper/ Zn-SOD is found in both chloroplasts and cytosols, whereas Mn-SOD is found in the matrix of mitochondria. Iron-SOD has been reported in chloroplasts, mitochondria, and peroxisomes of petals in a few plants. Superoxide dismutase has been one of the most widely studied enzymes in the antioxidant system. The importance of SOD for aerobic growth has been established by demonstration that the SOD-deficient mutant of *E. coli* are hypersensitive to oxygen. Recent research indicates that tobacco and alfalfa expressing high levels of Mn-SOD targeted to chloroplasts or mitochondria are more tolerant to paraquat and freezing, respectively.

Copper/Zn-SOD over-expression in tobacco increases tolerance to oxidative treatments. In the case of tobacco over-expressing Cu/Zn-SOD in the cytoplasm and chloroplasts, the plants are markedly more resistant to loss of photosynthetic capacity (photoinhibition) when exposed to high light intensity at low temperature. Several researchers have indicated that SODs play an important role in water stress tolerance of plants.

**Plant Antioxidant System**

Oxidative stress, characterised by excess accumulation of reactive oxygen radicals, could result from various environmental stresses such as water stress, chilling stress, air pollution, and herbicides. It was noted that antioxidant activity is significantly different between stress-sensitive and stress—tolerant cultivars of the same species when they are grown under stress. In other words, plant antioxidants are sensitive to stress, and their activity may be related to the capacity of plant stress tolerance.

*Alpha-tocopherol*: Photosynthetic apparatus and membrane could be affected by water stress. Alpha-tocopherol is a lipid-soluble antioxidant associated with biological membrane of cells, especially the membrane of photosynthetic apparatus. Research has shown that water-deficiency may result in an increase of tocopherol concentration in plant tissues. Burger et al. noted that tocopherol content of soybean leaves was increased as the amount of rainfall decreased. This is consistent with the reports of Tanaka et al., who showed that subjecting spinach to water deficit increased the content of $\alpha$-tocopherol in the leaves.

*Ascorbic acid*: Ascorbic acid is distributed in the cytosol as a water-soluble antioxidant. Dry conditions have been noted to increase the AA content in turnip greens, onions, and black currant. Black currant grown in hot, dry years were found to contain more than twice the AA of those grown in wet years. Stocker has reported an enhancement in AA content of plants which were subjected to severe water stress. Research results indicate that, under moderately dry conditions, a slight increase in the total AA concentration occurred.

There was an extraordinary upsurge in the total AA content along with abnormally high oxidation-reduction ratio and very high level of respiration under extreme drought conditions, suggesting a complete breakdown of the metabolism. Mukherjee and Choudhuri noted that AA increased significantly under water stress. In contrast, Price and Hendry found a decrease of ascorbate content of drought-stressed treatment in 9 out of 10 grass species. It is obvious that the relationship between antioxidant activity and drought stress tolerance is still not clear.

*Beta-carotene*: Beta-carotene can protect chlorophyll and prevent photoinhibition. It is closely associated with photosynthetic apparatus and quenches singlet oxygen efficiently. Stuhlfauth et al. found that acclimation of Digitalis lannta to low water potential results in a 25% increase of β-carotene.

*Superoxide dismutase*: Superoxide dismutases are closely related to the capability of plant tolerance to various stresses, such as water stress, chilling stress, herbicides, and pathogens. It was reported that SOD enhances water stress tolerance of plants. In tomato, cytosolic Cu/Zn-SOD was induced strongly by drought, while chloroplastic Cu/Zn-SOD remained largely unaffected. In two mosses, the drought-tolerant Tortula ruralis and drought-sensitive Cratoneuron filicinum were compared. The drought-tolerant moss showed lower levels of lipid peroxidation, together with increased levels of SOD; the opposite occurred in the sensitive moss.

Drought tolerant and intolerant maize inbreds were analysed by Malan et al., and resistance was found to correlate with Cu/Zn-SOD. Drought stressed tomato showed increases both in transcript levels of cytosolic Cu/Zn-SOD gene (eight-fold) and enzyme activity (five-fold), whereas an increase in enzyme activity was observed in chloroplastic Cu/Zn-SOD (2.4-fold) without a corresponding increase in transcript levels.

### Plant Growth Regulators and Antioxidant Activity

Both plant hormones and antioxidants are endogenous organic substances that regulate plant growth and development. They play an important role in plant tolerance to water stress. As water potential decreases, increased ABA and reduced cytokinin result in stomatal closure. GA and IAA are reduced due to the water stress. Ethylene increases rapidly as a response to water stress. The alteration of these hormones may facilitate plants to adapt to stressful environment and improve water stress tolerance by retaining moisture in the plant tissues. It has been noted that various antioxidants change rapidly as water potential decreases.

In most cases, α-tocopherol, β-carotene, and SODs are increased significantly under water stress. The change of the antioxidants may enhance water stress resistance of plants through detoxifying reactive oxygen species. The biochemical relationship between plant hormones and antioxidant responses to water stress is being revealed. It was noted that cytokinin can not only prevent the formation of reactive oxygen species, but also scavenge the free radicals. Therefore, cytokinin may play a dual role in suppressing free radical damage and water stress injury.

Seaweed extracts, as a cytokinin—containing material, may increase antioxidant activity and improve water stress resistance through suppressing reactive oxygen species. Water stress induces ethylence production. Hildebrand and Grayburn reported that the increase of ethylene under oxidative stress is associated with lipid peroxidation caused by activated oxygen radicals. Lipoxygenases are a class of dioxygenases that catalyse the peroxidation of polyunsaturated fatty acids and other molecules containing cis,cis-1,4-pentadiene moieties. Kacperska and Kukacka-Zebalska noted the similarities between lipoxygenase activity and ethylene production and proposed lipoxygenase as a possible in vivo ethylene—forming enzyme.

A series of studies were conducted to examine the antioxidant status, drought and disease tolerance, and growth response to foliar application of soluble seaweed (*Ascophyllum nodosum*) extracts (SE) and humic acid (HA; 25% active HA or 2.9% active HA) in tall fescue (*Festuca arundinacea* Schreb), Kentucky bluegrass (*Poa pratensis* L.) and creeping bentgrass (*Agrostis palusttis* Huds.) grown under low (-0.5 MPa) and high (-0.03 MPa) soil moisture environments. Some of these studies are briefly described in the following sections of this chapter.

## Effects of PGRs on Turfgrass Growth and Quality

Turfgrass growth and quality are significantly impacted by various biotic and abiotic stresses. Traditionally, a high quality turf is maintained through proper fertilisation, irrigation, mowing, disease control and other cultivation practices. However, as water resources decline, drought stress becomes a major limiting factor in turf management in many parts of the world. Water stress causes stomatal closure, reduces photosynthesis, and suppresses growth. Additionally, improper management of turfgrass, such as excess nitrogen application, results in increased disease occurrence and thus poor turf quality.

Utilisation of plant growth regulators has become an accepted practice in some turfgrass management systems. Traditionally, PGRs are used as growth retardants to suppress shoot growth and thus reduce mowing frequency. Most of these PGRs are synthesised inorganic or organic chemicals, and some of them are herbicides. The large amounts of these chemicals used during the growing season has increased environmental concerns. In the past decade, hormone-containing natural products, a group of PGRs, have received increasing interest. The PGSs not only improve turfgrass growth and stress resistance but also avoid environmental pollution.

Humic acid (HA) and seaweed extract (SE) are two plant growth regulators. Seaweed extract contains not only various mineral nutrients but also phytohormones and vitamins. When applied to plants, SE not only increases endogenous cytokinin concentration but also enhances nutrient uptake, increase shoot and root growth, increases chlorophyll content, and improves seed gemination. Seaweed extract has been shown to improve turfgrass growth and increase tolerance to environmental stresses. Humic acid contains valuable plant nutritional substances not found in chemical fertilisers.

Research showed that HA, which possess auxin-like activity, can be used as a growth regulator to regulate endogenous hormone levels, improve plant growth, and facilitate stress tolerance. Dollarspot, caused by *Sclerotinia homoeocarpa Bennett,* is a serious disease of turfgrass on golf course putting greens, closely mown fairways, bowling greens, and home lawns. Creeping bentgrass and annual bluegrass are highly susceptible to *S. homoeocarpa.* Dollarspot has traditionally been reduced with high rates of N fertilisation, adequate water supply, avoidance or removal of dew and guttation fluid from turf, and the application of fungicides.

No dollarspot-resistant cultivars of creeping bentgrass or annual bluegrass are available. Since disease-free golf greens and fairways are highly valued, frequent applications of inorganic or synthetic-organic fungicides are necessary for controlling this disease. However, large amounts of fungicides used each year could result in environmental pollution. Therefore, development and implementation of an alternative control for dollarspot may reduce fungicide use and minimise the risk of the pollution of water resources.

### Alpha-tocopherol Status and Growth Responses of Kentucky Bluegrass

Water deficits limit the growth and distribution of natural vegetation and the performance of cultivated plants more than any other environmental factors. Water stress causes various physiological and biological changes of plants. The decrease in photosynthesis, the closure of stomata, and osmotic adjustment are considered to be typical plant responses to the first stage of water stress. Closure of stomata by a decrease of water content in plant leaves disturbs the supply of $CO_2$ for photosynthesis.

Reduction of stomatal conductance and photosynthesis in water-stressed plants results in exposure of cells to excess excitation energy, especially under high-light conditions. The excess excitation energy may be diverted to activate molecular oxygen and stimulate excess production of active oxygen species which are toxic to cells. These active oxygen species such as superoxide, hydrogen peroxide, hydroxyl radicals, and singlet oxygen may damage proteins, lipids and nucleic acids and cause dysfunction of plant metabolism. Increasing evidence shows that many damaging environmental stresses, such as drought, directly or indirectly, occur through the formation of reactive oxygen species following impairment of the electron transport system.

Plants possess intrinsic antioxidant defence systems to detoxify reactive oxygen species and minimise oxidative injury. These include lipid soluble, membrane-associated antioxidants (e.g. α-tocopherol, β-carotene), water soluble reductants (e.g. ascorbic acid, glutathione), and enzymatic antioxidants (e.g. superoxide dismutase, catalase and the enzymes of the ascorbate/ glutathione cycle). It has been suggested that the formation of reactive oxygen species is an inherent consequence of metabolism and that control of their levels is essential for normal function of cells. Alpha-tocopherol (vitamin E), which is hydrophobic and associated with membranes, is one of the most important components of cellular defence against oxidative injury. It reacts with superoxide, lipid peroxy radical, and singlet oxygen and terminates the chain reactions of lipid peroxidation.

A significant increase of α-tocopherol concentration is observed in several plant species when they are subjected to drought. Burger et al. noted that α-tocopherol content of soybean leaves is increased as the amount of rainfall decreased. Tanaka et al. showed that subjecting spinach to water deficits increased the content of α-tocopherol in the leaves. Price and Hendry found that drought led to the production of superoxide radical anions in the plants, and, as the drought continued, there was a rise in the peroxidation of membranes and loss of control over the uptake of transition metals, especially that of Fe, into the cells.

Drought stress was also noted to cause an increase in the radical scavenging vitamin E (α-tocopherol) and an increase in glutathione in the plant tissues. Highly significant correlations were observed between stress tolerance and tocopherol concentration (Spearmans rank correlation coefficient r =0.731). Moran et al. noted that drought (leaf water potential = 1.3 MPa) induced a 67% increase of α-tocopherol. Proper application of plant growth regulators to turfgrasses can increase plant vigour and growth, stimulate root growth, and improve the resistance to various environmental stresses (drought, salinity, temperature, etc.).

Seaweed extracts (SE) improved cell membrane fluidity and permeability, resulting in better tolerance to drought in ryegrass. However, it is not clear whether this improvement is due to the effects of plant growth regulators on the status of reactive oxygen species scavengers (such as α-tocopherol). Little research has been done on the biochemical mechanisms of exogenous PGRs' effects on the antioxidant and water stress tolerance.

Some researchers done a study to determine the α-tocopherol concentration of Kentucky bluegrass. Methods and findings of their study are described below.

*PGRs application and plant culture*: Kentucky bluegrass (*Poa praetensis* L. cv. 'Plush') seeds were sown at 0.12 g in a 5 x 10-cm diam. plastic ring. Each ring contained the same amount of Groseclose silt loam soil with the ring bottom covered with plastic film. The dry SE was dissolved in water, and the SE solution was sprayed evenly on foliage in a rate of 80 ml $m^2$. The HA was diluted before application at a rate of 80 ml $m^2$.

*Drought stress treatment*: Drought stress treatment began on 3—wk old seedlings. Silt loam top soil was air-dried and then sieved through a 2-mm screen. This soil was mixed with appropriate water to obtain two moisture levels: 11.6% and 20.1% (which are equivalent to a water potential of—0.5 and—0.03 MPa, respectively). A 20-20-20 soluble fertiliser was mixed with water to provide 34 kg N $ha^{-1}$. After thoroughly mixing the soil with water, the soil was placed in a plastic bag, sealed and allowed to equilibrate for 72 hr. Terrarium-like boxes (30 x 40 x 75 cm) were constructed with 0.15 mm clear plastic sides and tops.

Into each box was placed 15 kg of soil adjusted to one of the two moisture levels. The plastic bottom of the rings were removed and the seedling in each ring was transplanted to the terrariums. Each of the terrariums received seven plugs. The seven treated grass plugs within each moisture level were arranged randomly within each terrarium and replicated three times. The terrariums tops were sealed with clear plastic film and the terrariums were randomly placed in an air-conditioned greenhouse at The Turfgrass Research Center of Virginia Tech, Blacksburg, Va, and rotated ¼ a turn three times a week. Thermocouples were used to record air temperature inside each terrarium.

Average air temperature was maintained at 23°C and 25°C at 8 am and 2 pm, respectively, during the experiment. Fresh leaves of the 6-wk old seedlings were sampled from each treated ring. The leaves were frozen with a small amount of liquid N, and then stored in -20°C before being homogenised. Plant height was measured and the plugs were removed from the growing medium. The soil was washed away from the roots. The root length was measured, and the root mass was dried in an oven at 60°C for 24 hours before being weighed.

*Extraction of α-tocopherol from leaves*: Alpha-tocopherol analysis was based on the methods of Tanaka et al.. Ethanol (80% aqueous solution) was used as the extraction medium for α-tocopherol. Fresh leaf segments (500 mg) of Kentucky bluegrass were homogenised with a Polytron PT 3000 (Kinematia AG) in 10 ml of 80% ethanol with 0.01% of butylated hydroxytoluence (BHT) at a speed of 29,200 rpm for 2 minutes. The homogenate was then centrifuged in the presence of Argon at 2,800 rpm for 30 min at 16°C. To 5 ml of the clear supernatant, 1.0 ml of n-hexane (85%) was added, and the mixture was fully agitated with S/P Vortex Mixer (Baxter Scientific products) for 2 minutes.

*Separation and determination of α-tocopherol*: Two hours after agitation of the mixture, the upper layer of hexane in the mixture was collected, filtered with a Nylon Acrodisc 13 mm x 0.2 μm filter and applied to a High-Performance Liquid Chromatography (HPLC). The HPLC system consisted of a HPLC pump, with a 20 μl injection loop, a variable wavelength fluorescence detector (Shimadzu RF 535 fluorescence HPLC monitor), a HPLC system controller, and a CR 501 chromatopac. Separations were performed on an analytical column of Supelcosil LC-Si (4.6 X 250 mm ) from Supelco and filtered with a guard column (50 x 4.6 mm) at room temperature.

The mobile phase was a mixture of hexane-isopropanol (99:1). Alpha-tocopherol was detected by measuring the fluorescence intensity at 325 nm while exciting at 292 nm. Alpha-tocopherol from Sigma was used to develop a standard curve for the calculation of α-tocopherol concentration in the samples. The α-tocopherol concentration was adjusted based on the recovery percentage (82%). The data were subjected to ANOVA analysis, and treatments were compared with Duncan's multiple range test.

### *Plant Growth Responses to Water Stress and PGRs*

The average leaf water status, in terms of water stress level (WSL) determined with a Campbell hydraulic leaf press, was decreased by 16.7% due to the drought stress. This showed that the water stress applied in this study was severe enough to significantly reduce leaf water content. Exogenous application of SE and/or HA improved the leaf water status significantly, especially under the water stress conditions. The highest leaf moisture content occurred when the seedlings were treated with HA and /or SE at the higher rates.

The shoot growth was suppressed substantially due to the water stress. On average, plant height and clipping dry weight of the grass grown under low soil moisture were only 38.7% and 21.5%, respectively, of those in the high soil moisture. However, exogenous PGR enhanced the shoot growth of Kentucky bluegrass under both low and high soil moisture conditions. The plant height and fresh clipping weight were significantly increased with application of the high doses of SE, HA, or their combinations.

Root length was not significantly influenced by soil moisture level or PGR treatments. The high dosage of HA, SE alone and the combination tended to have longer roots. The root dry weight was not significantly reduced under low soil moisture. Application of the high doses SE and/or HA increased root mass significantly in both low and high soil moisture environments. The PGRs appeared to have better effects on root development under water stress than under the high soil moisture conditions.

#### *Response of α-tocopherol*

Alpha-tocopherol concentration increased from 15.0 $\mu g g^{-1}$ dry leaves in the high soil moisture regime to 45.3 $\mu g g^{-1}$ dry leaves under water stress. Foliar application of the PGR further increased α-tocopherol content of Kentucky bluegrass significantly under both high and low soil moisture levels. The highest -tocopherol concentration occurred when the seedlings were treated with HA at 5 l $ha^{-1}$ or HA plus SE at 2.5 l + 326 g $ha^{-1}$. The results of this study indicated that exogenous SE or HA, two PGRs, enhanced endogenous concentration of α-tocopherol.

#### *Antioxidant α-tocopherol and Turfgrass Growth*

A positive correlation between α-tocopherol concentration and root or shoot growth was observed in Kentucky bluegrass under both high and low soil moisture. The correlation coefficients (r) were statistically significant ($p<0.1$) between α-tocopherol concentration and root dry weight or shoot dry weight. This indicated that plants endogenous concentration of antioxidant α-tocopherol is related to the plant growth.

### Effects of PGRs on Hydrophobic and Hydrophilic Antioxidant Status and Growth of Three Turfgrass Species

Water deficits are a major limiting factor in turfgrass management. Since turfgrass is grown under varying environmental situations, tolerance mechanisms of the plant are major strategies for turfgrass to survive water stress environments. There is wide variation among turfgrass species in terms of drought tolerance. It has been well documented that tall fescue shows better drought tolerance than Kentucky bluegrass or creeping bentgrass. Although the deep root system of tall fescue has been attributed to its good drought tolerance, the physiological basis of drought tolerance still remains unclear.

The antioxidant content of a plant is closely related to its stress tolerance. It has been established that environmental stresses may damage the plants through accumulation of toxic reactive oxygen species such as peroxide, hydrogen peroxide, hydroxyl radical, and singlet oxygen. Drought-induced oxidative damage has been observed in higher plants. The severity of this damage largely depends on the status of antioxidant systems, since plants develop antioxidants to remove toxic reactive oxygen species and protect the plant cells from lipid peroxidation and inactivation of enzymes that occur under stress.

Plant antioxidant status exhibits a wide variation from species to species, with a close relationship between antioxidant activity and water stress tolerance. Price and Hendry indicated that the antioxidant α-tocopherol content increased significantly as a response to water stress in 9 out of 10 grass species examined. Tanaka et al. showed that drought stress enhanced antioxidant activity in spinach. Several researchers reported that drought induced a significant increase in the antioxidant status in plants. Analysis of antioxidant status of grasses with different water stress tolerance may lead to a better understanding of the mechanisms involved.

Alpha-tocopherol and ascorbic acid are two important antioxidants in higher plants. They are concentrated in the chloroplasts and the cytosol (ascorbic acid), and protect the photosynthetic apparatus under stress by scavenging excess reactive oxygen species. Although some research on the α-tocopherol and ascorbic acid responses to drought has been reported, the manipulation of endogenous α-tocopherol and ascorbic acid in drought stressed turfgrass with exogenous growth regulators has not been well established. Regulation of turfgrass performance with plant growth regulators has been practiced for many years.

Proper application of certain PGRs can not only improve turfgrass growth but also enhance stress tolerance. Seaweed and HA are two environmentally compatible natural products. Containing phytohormones and showing auxin, or cytokinin-like activity, when these materials are applied to turfgrass, they enhance the tolerance to salinity, drought, chilling and other environmental stresses. Examination of different stress tolerant grasses treated with these plant growth regulators may provide information for managing turfgrass for better tolerance to water stress.

Some researchers done a study to examine the influence of exogenous plant growth regulators on endogenous hydrophobic and hydrophilic antioxidant concentrations and growth of three turfgrass species. The methods and finding of their study are described below.

### PGRs Application and Drought Stress Treatment

Creeping bentgrass, Kentucky bluegrass (*Poa praetensis L. cv. 'Plush'*), and tall fescue (*Festuca arundinacea Schreb. cv. "Rebel Jr."*) seeds were sown at rates 0.06, 0.12, and 0.25 g ring$^{-1}$, respectively, in a 10-cm diameter ring. Each ring had the bottom covered with plastic film and contained the same amount of Groseclose silt loam soil.

When the seedlings were 2-wk old, soluble seaweed (*Ascophyllum nodosum*) extracts (SE) at a rate of 326 g $ha^{-1}$, humic acid (HA; 25% active humic acid) at a rate of 5 l $ha^{-1}$ or the combination were sprayed evenly on the foliage. Silt loam top soil was air-dried for 2 weeks and then sieved through a 2-mm screen. This soil was mixed with appropriate water to obtain two moisture levels: 11.6% and 20.1% (which are equivalent to a water potential of $-0.5$ and $-0.03$ MPa).

A 20-20-20 soluble fertiliser was mixed with water to provide 34 kg N $ha^{-1}$ (3 g N $box^{-1}$). After thoroughly mixing the soil with water, the soil was placed in a plastic bag, sealed, and allowed to equilibrate for 72 hr. Terrarium-like boxes (30 x 40 x 75 cm) were constructed with 0.15 mm clear plastic sides and tops. Fifteen kg of soil adjusted to one of the two moisture levels was placed into each terrarium. One week after treatment of the seedlings, the plastic bottom of the rings were removed and the seedling in each ring was transplanted to the terrarium. Each of the terrariums received 8 plugs.

The terrarium tops were sealed with clear plastic, and the terrariums were placed in an airconditioned greenhouse. Thermocouples were used to record air temperature inside each terrarium twice daily. The average temperature was 23°C at 8 am and 26°C at 2 pm, respectively, during the period of the experiment. Water stress level (WSL) for all treatments were measured with a hydraulic leaf press 7 wk. after germination. In addition, plant height was measured and fresh leaves of the 7-wk old seedlings were sampled from each treatment at this time.

The leaves were frozen with a small amount of liquid N, and then stored at -20°C before being homogenised. Following the WSL measurement, the soil was washed from the roots which were dried in an oven at 60°C for 24 hrs before being weighed. The four PGR treatments within each moisture level were arranged randomly within each terrarium and replicated four times.

### Extraction of α-tocopherol from Leaves

The fresh leaves were sampled and frozen with liquid N and stored at -20°C. The frozen leaves segments (1,000 mg) were homogenised in 7 ml of 5% metaphosphoric acid with a Polytron (PT3000, Kinematia, AG) at a speed of 29200 rpm for 3 minutes. Ice water was used to maintain the samples at a low temperature during the homogenisation. The homogenate was then centrifuged with a refrigerated centrifuge at 4000 rpm for 40 minutes.

Clear supernatant (4 ml) was analysed by a High-performance Liquid Chromatography (HPLC). Determination of α-tocopherol concentration. The fully agitated mixture stood for 2-hr before further analysis. The α-tocopherol concentration were adjusted based on the recovery percentage (82%). A two factor analysis of variance (ANOVA) and a least significant difference (LSD) test were performed on each data set.

### Growth Responses to Water Stress and PGRs

The average leaf water status, in terms of water stress level (WSL) determined with the hydraulic press, was decreased due to the water stress in the three cool-season turfgrass species. Among the three grasses, Kentucky bluegrass appeared to have the best water status (the lowest WSL). The leaf water status was improved significantly with application of the PGRs under both high and low soil moisture environments.

In other words, SE and/or HA enhanced water stress tolerance. It appeared that the combination of SE and HA had better effects on plant water status than when either material was applied alone. Water stress reduced shoot growth in the three turfgrasses examined. The plant height and clipping dry weight were significantly reduced under drought stress in the three grass species. On average, shoot dry weight was reduced by 25.7% in creeping bentgrass, 38.6% in tall fescue, and 21.4% in Kentucky bluegrass, respectively, under drought as compared with those in high soil moisture levels. Application of the PGRs enhanced plant height of the three grasses under both high and low soil moisture.

The clipping dry weights were increased with application of PGRs to the three grasses grown under the high and low soil moisture environments. It appears that SE and/or HA have greater stimulating effects under water stress than under high soil moisture. Among the three grasses, creeping bentgrass showed the greatest response to the PGRs. Application of the PGRs improved root growth as determined by root length measurements in three grasses grown under both the high and low soil moisture. However, only the combination of HA and SE enhanced root length significantly in the three grasses. The root dry weight of tall fescue and Kentucky bluegrass was significantly enhanced with all PGR treatments.

Although application of HA or SE alone did not increase root mass of creeping bentgrass, the combination of HA and SE caused a significant increase of root mass of creeping bentgrass. The interaction between growth regulators and moisture levels differed with grass species. The moisture and PGR interaction for shoot weight was significant in creeping bentgrass, but not in the other two species. In contrast, the moisture and PGR interaction for root length was significantly in tall fescue and Kentucky bluegrass but not in creeping bentgrass. The SE and HA showed greater stimulating effects on root growth under drought than in the high moisture conditions.

### *Response of α-tocopherol and Ascorbic Acid*

Water stress resulted in a significant increase of exogenous α-tocopherol concentrations in the three turfgrass species. The combination of SE with HA enhanced α-tocopherol content in the three grasses under high and low moisture. Among the three grasses, tall fescue and Kentucky bluegrass contained a higher level of α-tocopherol and a lower level of ascorbic acid as compared with creeping bentgrass. Water deficit did not influence ascorbic acid content in all three grass species. Application of the PGRs significantly enhanced ascorbic acid content in the three grass species. This showed that a lipid-soluble, membrane-associated antioxidant (α-tocopherol) may be more sensitive to water deficit than a water-soluble reductant (ascorbic acid).

However, the ascorbic acid concentration on dry weight basis was enhanced with application of the PGRs. Since the experimental design did not include the grass species, no statistical comparison can be done on the ascorbic acid content among the three grasses. However, because the three grasses were treated on the same date and grown under the same experimental conditions and antioxidant analysis procedures were the same, the ascorbic acid content of three grasses was compared based on the average of different PGR treatments. The data showed that the ascorbic acid status of tall fescue and Kentucky bluegrass exhibited greater response to the PGR treatments than did that of creeping bentgrass.

### *Antioxidant Status and Turfgrass Growth*

Positive correlations between antioxidant (α-tocopherol

and ascorbic acid) concentrations and root or shoot growth were observed in the three grasses grown under both high and low soil moisture levels. Since the relationships between the antioxidants and plant growth were similar in both tall fescue and Kentucky bluegrass, the comparisons were made only for tall fescue and creeping bentgrass.

However, the correlation between α-tocopherol and shoot or root weight differed with grass species. Tall fescue contained a higher level of α-tocopherol and showed a closer relationship between the antioxidant and growth (the r value ranged from 0.8936 to 0.9637) as compared with creeping bentgrass ( the r value ranged from 0.6496 to 9479). Similarly, there was a positive correlation between ascorbic acid and shoot or root growth in tall fescue and creeping bentgrass grown under drought and high soil moisture environments.

A significant correlation between α-tocopherol and ascorbic acid content was found in tall fescue and creeping bentgrass grown under low soil moisture condition, but not in creeping bentgrass under high soil moisture. This result indicated that the turfgrass with high level of α-tocopherol generally contained more ascorbic acid.

### Antioxidant Status and the Drought Tolerence of Kentucky Bluegrass

Water deficit stress affects plants at various levels. Stomatal closure, reduction of photosynthesis and osmotic adjustment are typical responses of plants to water stress at the first stage. Water stress-induced stomatal closure interferes with exchange of $CO_2$ and $O_2$ in the cell. It is well documented that a large proportion of the electron flux is diverted from $CO_2$ assimilation to

$O_2$ reduction under stressed conditions as a consequence of the decrease in Ci (intercellular $CO_2$ concentration) and $g_s$ (stomatal conductance).

Reduction of $O_2$ in the Mehler reaction results in formation of reactive oxygen species such as superoxide and hydrogen peroxide. In addition, absorption of excess light by chlorophyll leads to the formation of triplet excited chlorophyll which can then pass excitation energy to oxygen resulting in formation of singlet oxygen. Research results suggest that many environmental stresses, such as drought, salinity, low temperature and herbicide pollution, damage plants through the reactive oxygen species.

Plants develop various antioxidants to cope with oxidative stress. Alphatocopherol, β-carotene, and superoxide dismutase (SOD) are important antioxidants which protect plants by suppressing oxidative injury. Alpha-tocopherol, which is located within biological membranes, can quench oxygen radicals, stabilise cell membranes by influencing lipid organisation, and protect chlorophyll. Water deficit may result in an increase of tocopherol concentration in plant tissues. Based on the studies of 10 different grass species under water stress, Price and Hendry found that drought stress led to an increase of one to three fold of α-tocopherol concentration in 9 out of 10 species. They pointed out that species with a high tolerance of stress are defended through tocopherol.

Beta-carotene can protect chlorophyll and prevent the plants from photoinhibition. It is closely associated with the photosynthetic apparatus and quenches singlet oxygen efficiently. Stuhlfauth et al. found that acclimation of Digitalis lannta to low water potential resulted in a 25% increase of β-carotene. Superoxide dismutase is the most efficient scavengers of the superoxide anion and an essential component of the ascorbate-glutathione cycle for the detoxification of toxic reactive oxygen species.

The close relationship between SOD activity and drought stress tolerance has been observed in several crops, such as maize and tobacco. In tomato, cytosolic Cu/ZnSOD was induced strongly by drought while chloroplastic Cu/ZnSOD remained largely unaffected. Drought stressed tomato showed increases both in transcript levels of cytosolic Cu/Zn SOD gene (eight fold) and enzyme activity (five fold); whereas, an increase in enzyme activity was observed in chloroplastic Cu/Zn-SOD (2.4-fold) without a corresponding increase in transcript levels. Although it was reported that major antioxidants showed differential responses to environmental stress, few studies have been done on the relationship among different antioxidants under drought stress.

It has been proposed that α-tocopherol is maintained until ascorbic acid has been completely oxidised, then α-tocopherol depletion occurs. Kunert and Ederer examined the relationship between vitamin E (α-tocopherol) and vitamin C (ascorbic acid) and found that the vitamin C to vitamin E ratio changes during the life cycle of the beech leaf. Finckh and Kunert noted that both the amount and ratio of vitamin E and C contributed to the sensitivity of a seedling to the herbicide diphenylether.

Proper application of certain plant growth regulators may improve turf quality and enhance stress tolerance. Both SE and HA, which possess cytokinin-like and auxin-like properties, can enhance plant growth, delay senescence and improve turf quality, increase tolerance to drought, and improve nutrient uptake. However, the physiological mechanisms of the PGRs' effects on drought tolerance is still not fully understood.

Some researchers done a study in Kentucky bluegrass to understand the antioxidant status and the drought tolerence. The methods and findings of this study are briefly described below.

#### *Plant Culture, PGRs Application, and Drought Stress Treatment*

Seeds of Kentucky bluegrass (*Poa pratensis L. cv. 'Plush'*) were sown in a 10-cm diameter and 5 cm deep ring containing the same amount of Groseclose silt loam soil at 0.12 g $ring^{-1}$ in the greenhouse on 11 November 1995. Water stress treatments began when the seedlings were 3-wk old. The terrarium-like boxes (30 x 40 x 75 cm) were used, and two moisture levels (11.6 % and 20.1% which were equivalent to a water potentials of -0.5 and -0.03 MPa, respectively) were obtained by adding appropriate water. The treated seedlings of Kentucky bluegrass were permitted to grow in the terrariums under one of the two moisture regimes for 5 weeks. The fresh leaves from each treatment were sampled and frozen immediately with liquid N and then stored at - 20°C.

#### *Determination of Antioxidants*

*Beta-carotene*: Beta-carotene was extracted with 40 ml of HPLC-grade hexane containing 0.01% butylated hydroxytoluene (BHT). Frozen leaf segments (500 mg) were homogenised with a Polytron PT3000 (Kinematia AG) in the extraction medium for 2 minutes. The crude extract was washed with a saturated NaCl solution. Trace water was removed from hexane by filtering through anhydrous powdered sodium sulphate (5 g). The hexane phase was then evaporated to dryness and redissolved in a mobile phase containing MeOH/Acetonitrile/DCM (40:48:12, v/v/v).

The sample in the mobile phase was filtered with a Nylon acrodisc 13 mm x 0.2 μm filter and analysed by a

HPLC. The HPLC system consisted of a HPLC pump, with a 20 μl injection loop, a UV detector (SPD-6A), a HPLC system controller (Shimadzu SCL-6B), and a CR501 chromatopac. Separations were performed on an analytical column of Pinnacle 3 μm ODS (100 x 4.6 mm) from Restek Corporation with a C-18 guard column (50 x 2.0 mm) ( Whatman 30-38 μm mesh). The mobile phase was a mixture of methanol/acetonitrile/dichloromethane (40:48:12, v/v/v).

Beta-carotene was detected with the UV detector set at 450 nm, and the concentration of β-carotene in the plant extracts was determined using peak height against the standard curve. Light was eliminated with aluminum foil, and dilutions were prepared when necessary to keep the absorption in the linear range of the standard curve. The β-carotene concentrations were calculated based on the standard curve.

*Superoxide dismutase*: Superoxide dismutase was analysed according to the methods of Giannopolitis and Ries with the following modifications. The frozen samples (1.0 g fresh weight) were homogenised with a Polytron in 10 ml of 0.05 M $Na_2HPO_4/NaH_2PO_4$ (pH 7.0) buffer. The homogenates were filtered through four layers of cheesecloth and then centrifuged at 4°C for 20 min at 15,000 x g. The supernatants were collected and used for SOD assay. The crude SOD extracts were diluted ten fold with the extraction medium. The reaction mixture was composed of 1.3 μM riboflavin, 13 mM methionine, 63 μM nitro blue tetrazolium (NBT), 0.05 M sodium carbonate (pH 10.2), and appropriate volume (50 μl) of the diluted extract.

Deionised distilled $H_2O$ was added to bring to the final volume of 3 ml. The riboflavin was added last. The mixture was illuminated in glass test tubes selected for uniform thickness and colour. Identical solutions that were not illuminated served as blanks. The reaction device consisted of a glass test tube holder (two beakers) immersed in water maintained at 25°C. A circular fluorescent lamp (GE, FC 8T9-WW-RS) was attached on the outside of the beaker, and the whole device was set in an insulated chest. The reaction was initiated and terminated by turning the light on and off.

The light was on for 10 min. Appropriate dilution of the supernatant was performed when necessary to ensure that the absorbance value of the sample fell within the linear section of the standard curve. One enzyme unit of SOD activity is defined as the amount of enzyme required to cause 50% inhibition of nitro blue tetrazolium (NBT) reduction measured at 560 nm on a spectrophotometer. The SOD activity was determined according to the equation:

$$\text{SOD units g}^{-1}\text{ fw} = [(V/v) - 1]\text{ (dilution factor)}$$

Where $V = V_2 - V_1$, $V_1$ and $V_2$ are absorbance in absence of SOD without and with light, respectively. $V_1 = 0$. $v = v_2 - v_1$. $v_1$ and $v_2$ are absorbance in presence of SOD without and with light, respectively. The standard SOD (Sigma)(1 mg of the product contains 4400 units SOD) was used to develop standard curve to verify the relationship between SOD activity and V/v ratio at 560 nm.

#### *Response of Alpha-tocopherol to Water Stress and PGRs*

Water stress did not significantly influence α-tocopherol content (dry weight basis) in Kentucky bluegrass. However, application of SE and HA significantly enhanced α-tocopherol content under both low and high soil moisture.

#### *Response of Beta-carotene*

Water stress resulted in an increase of β-carotene in Kentucky bluegrass. Beta-carotene concentration was 272.9 and 420.3 μg $g^{-1}$ DM in the grass grown in high and low soil moisture, respectively. Application of HA plus SE enhanced β-carotene status under both low and high water stress conditions. β-carotene and α-tocopherol exhibited a similar response to exogenous application of the growth regulators.

#### *Response of Superoxide Dismutase*

The enzymatic antioxidant SOD activity of Kentucky bluegrass was not significantly influenced by water stress. However, exogenous application of SE and HA promoted SOD activity, especially under drought stress.

# 5

# Plant Growth Regulators for Cotton

Cotton plants produce several natural growth regulators or plant hormones, which modify plant growth and divert energy allocation within the plant. Producers often apply growth regulators to the foliage in an effort to maintain a balance of vegetative and reproductive growth. Plant growth regulators decrease cotton vegetative growth by modifying the production of plant hormones such as gibberellins, auxins and cytokinins. The most commonly used growth regulator is mepiquat chloride, which decreases vegetative growth by reducing gibberellic acid formation. Understanding the causes of excessive vegetative growth and the mode of action of Mepiquatbased growth regulators is essential for satisfactory vegetative growth management.

## Excessive Vegetation Growth

No optimal plant size exists for cotton; however, excessive vegetative growth can lead to several production problems. Fruit abortion, delayed maturity, boll rot, and harvest difficulties often coincide with excessive vegetative growth.

### Fruit Abortion

The cotton plant develops in an ordered fashion. Fruit initiates at the bottom of the plant and then proceeds up the plant and out to more distal fruiting positions as the season progresses. Excessive vegetative growth causes shading in the lower canopy and ma: lead to the abscission of early fruit (Figure. 1).

*Figure 1. Excessive shading can cause fruit shed or abscission A) Cotton plant with excessive fruit shed B) Cotton plant with normal fruit shed*

Numerous environmental and management factors can also contribute significantly to fruit shed, such as low light, excessive nitrogen fertility, and root pruning from cultivation.

### Delayed Maturity

Loss of early fruit may be compensated for when favourable conditions exist to extend the growing season. This compensatory growth results in plants setting more fruit at higher main-stem nodes and more distal fruiting positions than would occur normally. Compensatory growth generally results in delayed crop maturity. Boll size is strongly correlated with boll location on the plant.

(Figure 2) Bolls located on main-stem nodes 10-16 at the first or second fruiting position tend to produce the majority of crop yield.

If the cotton plant compensates for early season fruit loss by setting bolls later in the season, yield may still be compromised if water, nutrient, or temperature stress is encountered. Under these conditions, fruit produced outside this zone tend to be smaller. Compensatory growth may result in similar boll numbers but yield may still be reduced due to smaller boll size.

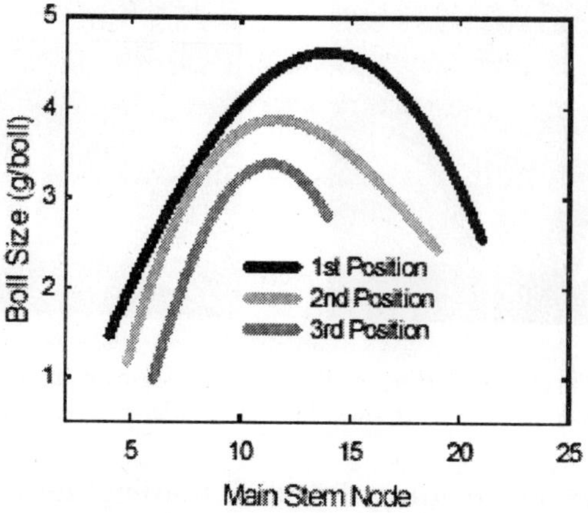


*Figure 2. Boll size is correlated to position on fruiting branch.*

### *Excessive Shading*

Excessive shading may also decrease the micronaire of bolls that are retained in the lower portion of the plant. If shedding does not occur boll rot is often more prevalent with excessive vegetation due to a more humid microenvironment in the lower portion of the canopy. Excessive vegetative growth inhibits penetration of insecticides to the lower portion of the canopy and can also reduce harvest of the crop. A smaller plant with a more compact fruiting zone is harvested more easily than a taller plant with excessive vegetative growth.

### Growth Hormones in Cotton Plants

The three types of growth hormones affected by foliar-applied products are gibberellins, cytokinins, and auxins. These hormones are integral for cotton growth and each has a unique role in the regulation of vegetative and reproductive growth.

Gibberellins, or gibberellic acid, promote cell division and expansion. This hormone is most closely related to vegetative growth. Mepiquat chloride, a foliar-applied growth regulator, reduces the concentration of gibberellic acid in the plant. When the concentration of this acid is reduced, a more compact plant cell, and subsequently a more compact plant develop.

Cytokinins promote cell division and elongation, but are also involved with the transport, accumulation, and retention of carbohydrates. While this hormone may contribute to the regulation of vegetative growth, it may also play a significant role in fruit retention and boll fill. Auxin increases cell wall plasticity and is considered a growth promoting hormone. Auxin also delays sensecence. This hormone is similar to gibberellic acid in its regulation of vegetative growth.

### Natural Growth Regulators

Several naturally occurring hormones work in the cotton plant to adjust plant growth. When plant growth regulators are applied to the cotton plant, they work in much the same way as the natural regulants already present. In many ways, they supplement or destroy the natural hormone. They often will work together in ratios and concentrations to regulate growth. Some naturally occurring hormones are auxins, gibberellins and cytokinins. Some of the more common plant growth regulators are composed of these hormones or combinations of them. The auxin family serves a number of functions in plants, such as cell and stem elongation, leaf expansion, increased rooting and changes in fruit abscission.

Some yield response has been noted when certain auxins have been applied to cotton, probably due to greater boll retention. Other reports have shown no yield response. The gibberellins are associated with stem elongation and some leaf enlargement, but have been shown to increase fruit retention in cotton. Yield responses when applied alone to cotton have been mixed, with some data showing smaller bolls and no yield response. Cytokinins can promote cell division and enlargement of stems, leaves and fruit. They have also been shown to reduce fruit abscission in cotton, but this is probably an indirect result of making the fruit a better competitor for nutrients. Cytokinins can also delay aging of leaves, allowing longer photosynthetic activity.

### Synthetic Growth Regulators

The number of synthetic plant growth regulators available to cotton growers is increasing. Pix has been the most widely used material due to its ability to reduce excessive cotton growth and put more emphasis on boll development. Mepichlor has more recently entered the market with the same active ingredient as Pix, Mepiquat Chloride. PGR IV is the first labelled plant growth regulator for cotton that promotes cotton growth. Maxxon has most recently followed in the same market track as PGR IV, with the same hormones. Cytokin has also entered the market with a different hormone than PGR IV or Maxxon.

### *Pix*

Pix (Mepiquat Chloride) is the first plant growth regulator in cotton production to make a significant impact on cotton growth and yield. Since its introduction by BASF Corporation in the early 1980s, it has been widely studied and frequently used in the major cotton growing areas. Pix is an anti-gibberellin, meaning that it reduces the production of gibberellin in the plant, which normally would enlarge plant cells. Mepichlor is produced by Micro Flo and contains the same active ingredient as Pix; therefore it is used in much the same way.

Pix applications are most successful when applied near the early bloom stage of development and again within 3 to 4 weeks on cotton that is subject to rapid and excessive vegetative growth. Not all cotton fields will benefit from Pix applications. It works best on healthy, nonstressed cotton when conditions favour rank growth. Many times, these conditions occur when water through irrigation and rainfall has been abundant, high nitrogen rates have been used, and large, indeterminate varieties have been planted.

The primary effect that Pix has on cotton production is that it takes energy away from leaf and stem development and directs it toward boll development and retention. For this reason, it is crucial for cotton growers to understand their individual field conditions and make or omit applications as conditions dictate. As a result of a Pix application at the appropriate times of development, lint is more efficiently produced due to decreased plant height, branch length and leaf size. This provides for a more open canopy that increases boll retention and reduces boll rot; consequently, bolls are given greater potential to produce. However, yield response will depend on the necessity of Pix in a particular field.

*Pix management*

Several approaches to using Pix have been developed across the Cotton Belt. All have credibility and can be successful in reducing growth, yet a universal approach that is used on every farm has been slow in coming. Conditions vary widely throughout the cotton growing areas. The most commonly used approach is to base application decisions on plant height at early bloom. The Pix label specifies that cotton should be between 20 and 30 inches tall near this growth stage before the first application is made.

Usually, Pix applied at the 1/2 to 1 pint per acre rate is recommended, followed 3 to 4 weeks later by a second application if necessary. Growers should remember that the label restricts the total amount applied in one season to 1-1/2 pints per acre. Following responses to different situations at early bloom:

Situation 1: Plant height is less than 20 inches at early bloom due to stress.

Response: Relieve stress if possible. Avoid a Pix application. Treatment may be warranted later.

Situation 2: Plant height is 20 to 24 inches at early bloom.

Response: If bloom begins before July 5, then the crop is on schedule. Pix at 1 pint per acre may be required later, particularly if plant size exceeds 28 inches within one week of early bloom or 32 inches within two weeks of early bloom. If bloom begins after July 5, particularly after July 10, then apply 1/2 to 1 pint per acre to shorten the boll loading period if crop is not under a drought stress.

Situation 3: Plant height is more than 24 inches at early bloom. Growing rapidly.

Response: Apply 1/2 to 1 pint per acre to reduce shading and improve boll set. An additional 1/2 to 1 pint (depending on previous treatment rate) may be required if the plant size exceeds 28 inches one week after early bloom or 32 inches two weeks after early bloom.

Situation 4: Plant height is approaching 20 to 24 inches prior to early bloom. Growth is rapid and the crop is well watered. Anticipated early bloom height is more than 24 inches.

Response: If prebloom cotton is 16 inches tall, apply 1/4 pint per acre. If prebloom cotton is 20 inches tall prior to first treatment, apply 1/2 pint per acre. An additional Pix treatment at 1/2 to 1 pint per acre may be necessary if the plant height exceeds 24 inches at early bloom, 28 inches after early bloom, or 32 inches two weeks after early bloom.

### *Low rate multiple application*

A drawback to applying these recommended rates in one or two applications is the possibility of drought stress that may set in following an application, creating an adverse effect on the crop. In response to this concern, a low-rate strategy has been developed. This approach uses low-rate multiple (LRM) applications of Pix beginning at squaring and continuing as crop growth dictates. The difference is a more conservative approach to slowing crop growth where applications may be ceased during conditions of stress. If growth is naturally deterred prematurely, growers have less of an investment tied up in the crop than if they used higher rates. Using LRM applications, up to four low-rate applications may be made to the crop with rates of 1/8 to 1/4 pints per acre.

The first application should be made at match head square and subsequent treatments are made at 7- to 14-day intervals or when regrowth occurs. Crop growth and growing conditions will dictate application rates and timing. As long as favourable conditions occur, this is an acceptable measure for controlling cotton growth. Care should be taken to scout the crop to ensure that excessive growth does not require a more aggressive approach.

*Plant monitoring approach*

A third approach to Pix management is currently being recommended in North Carolina and Arkansas. This approach is recommended for growers who are willing to employ plant mapping techniques in their management plan. It takes a systematic approach that is based on accumulated plant growth and the current rate of growth. Desired plant height and the assumed number of nodes at the end of the season are key ingredients in this recommendation. Several physiological assumptions must be made that are applicable to cotton growth in a particular area.

For Missouri, irrigated cotton should produce around 20 to 22 nodes during the season, while nonirrigated cotton may produce 17 to 18. These values may vary from year to year and field to field, yet they are still a good standard on which to base a decision. Growers using this system should establish final plant height goal. Under irrigated conditions, some threat of boll rot exists in Missouri conditions. For this reason, we suggest establishing a final plant height goal 2 inches greater than the row spacing. However, as a general recommendation that takes nonirrigated conditions into account, the following guidelines may apply.

When on 30-inch rows, it is suggested that 30 to 35 inches is optimum. Those growers using 38-inch rows typically should strive for 40 to 45 inches. Once a goal has been set for crop growth, a Pix threshold is derived by dividing the desired plant height by the expected number of nodes developed in the growing season. Next, during the first square stages or shortly after, a measurement is taken on plant height and nodes are counted on a representative sample obtained from a field. A grower should take five plants from 10 locations in the field to get an accurate measure. Again divide the plant height at the stage of sampling by the number of nodes. Compare this number with the Pix threshold.

If the number obtained during the measurement is less than the Pix threshold, Pix is not required. However, if it is greater than the threshold, further monitoring should be made. Since elongation is most pronounced in the top five nodes, measure the total length of the internodes between these nodes. Divide the inches of this active internode growth by three. A Pix application is suggested if the number exceeds the Pix threshold. Pix should be applied at lower rates similar to the LRM approach if a treatment is warranted around match head square.

As the plant develops into the early bloom stage, 1/2 to 1 pint may be needed to slow growth sufficiently. Remember, no more than 1-1/2 pints should be applied in one growing season, and the last application should be made no later than 30 days before harvest. The approach that Missouri growers opt to use depends on the level of management they are willing to expend to reach a goal.

We suggest that growers choose the method that they are most comfortable with.

One Mississippi study comparing LRM and single applications found primarily physiological differences versus yield responses to the different treatments. These researchers found that the early LRM approach made a greater impact on early season height, node number and internode length. The single early bloom application mostly impacted boll retention. A more effective height control is gained when applications are started early. However, the effect the application will make on yield is the bottom line.

### *Mepiquat Chloride Treatment*

Limiting vegetative growth in cotton has been and continues to be researched extensively. In the late-1960s and early-1970s, the chemical 2-chlroethyltrimethyl-ammonium chloride was used in India where excessive vegetative growth was a problem. Currently, mepiquat is used by many cotton producers to manage vegetative growth. Mepiquat inhibits a key enzyme in the production of gibberellic acid. The effect of mepiquat is a function of plant size and rate of product applied, and only affects new growth. Mepiquat has no effect on portions of the plant that have ceased growing. This means that a given rate of product applied at first-square to a relatively small plant has a much greater effect than that same rate applied to a larger plant in the bloom stage.

The morphological response of a cotton plant treated with mepiquat chloride is the suppression of vegetative growth through a reduction in the total number of mainstem nodes, and internode lengths. A reduction in leaf area is also observed with the use of mepiquat chloride. Since mepiquat chloride does not stop the growth of the cotton plant, the reduction in leaf size is also generally correlated with an increase in leaf thickness. Mepiquat chloride-treated plants are typically shorter, more compact, and possess a characteristically darker green colour than untreated plants. The use of mepiquat chloride may also cause a shift in boll location creating a more concentrated fruiting zone.

In a study by Kerby et al., both the mepiquat chloride treated and untreated cotton had similar numbers of bolls, yet the mepiquat chloride treated cotton had a greater percentage of those bolls on lower main stem nodes. It is important to realise that even without a mepiquat chloride treatment; boll retention on lower nodes can limit vegetative growth. High early fruit retention provides the plant with a reproductive sink limiting the resources available for vegetative growth.

### *Benefits of Mepiquat Chloride Treatments*

The benefits gained with a mepiquat chloride application are directly related to the morphological response exhibited by the treated plants. The size of a cotton plant can influence harvest efficiency. Taller and ranker plants significantly reduce the speed in which the crop can be harvested. Bolls set over a more compact zone on the plant increases harvest efficiency by allowing plant material to travel more easily through the picker head. Reducing crop height is the most consistent response from a mepiquat chloride treatment. The leaf area index (LAI) of a mepiquat chloride treated crop can be decreased by 5 to 10 percent compared to a non-treated crop. This lower LAI can improve pesticide distribution into the lower canopy.

Lower LAI also allows more light to penetrate the lower canopy for boll development. Most photosynthetic energy for boll development originates from adjacent or subtending leaves. Boll development in the lower portions of the plant on interior fruiting positions in shaded conditions may result in reduced maturity, quality, and potentially reduced yield. In extreme cases, shading of the lower canopy may cause fruit abortion. Despite many benefits of a mepiquat chloride application, yield effects with mepiquat chloride applications have always been inconsistent.

Studies have shown yields may be increased, decreased, or unaffected treatment. Yield effects are generally small when the cotton plant has high early fruit retention. Positive yield effects are more likely to occur when early fruit retention is reduced and vegetative growth is excessive. Increases in yield are usually 100 lbs lint/A, or less (Table 1). Yield reductions are more likely to occur with the use of mepiquat chloride at excessive rates in the presence of early-season environmental or management stress. Under these conditions the growth regulating effects of mepiquat act as an additional stress on the plant. Mepiquat chloride should be viewed as a management tool, not a yield enhancer.

*Table 1. Effect of mepiquat chloride treatment on yield vs. untreated check in large plot trials.*

| *Treatment* | *Rate* 2002 | 2003 | *Yield* |
|---|---|---|---|
| | *– oz/A –* | | *– lbs lint/A –* |
| Mepiquat pentaborate | 24 | 30 | 2962 |
| Mepiquat chloride | 24 | 30 | 2923 |
| Untreated | – | – | 2674 |
| Pr>f | | | 0.0009 |
| C.V. | | | 4.2 |

### *Types of Mepiquat Products*

#### *Mepiquat chloride*

Mepiquat chloride is the primary growth regulator used in cotton production. It is most commonly formulated as a 4.2 percent concentration of nitrogen, Ndimethyl piperidinium (or mepiquat) chloride salt. Mepiquat chloride is generally absorbed by the plant in 4 to 8 hours. The addition of a surfacetant may enhance uptake but is generally unnecessary. Mepiquat Chloride has been labelled for the use in cotton since the 1980's, and is sold under many trade names such as Pix, Mepex, Mepichlor, and Mepiquat Chloride.

#### *Mepiquat pentaborate*

Mepiquat pentaborate has been recently developed and is sold under the brand name Pentia by BASF Corporation. Pentia contains a 9.6 percent concentration of mepiquat as a pentaborate salt. While Pentia appears to have more active ingredient than the original mepiquat chloride formulations, this increase is due to the weight of the pentaborate salt, which is heavier than the chloride salt of mepiquat chloride. There is no difference in the amount of actual mepiquat between the two products. BASF has indicated the uptake rate of the pentaborate formula is greater than mepiquat chloride and is advertised as a more rain-fast product.

#### *Mepiquat chloride premixes*

As the use of Mepiquat chloride gained acceptance in the cotton industry, attempts have been made to improve the effectiveness and yield response to this product. Several "premixes" have been developed combining Mepiquat chloride with reproductive growth enhancers or other growth regulating hormones. Pix Plus contains a 4.2 percent concentration of Mepiquat chloride with the bacterium *Bacillus cereus*. This bacterium was touted to be a reproductive growth enhancer. Mepex Ginout also includes a Mepiquat chloride with kinetin, a synthetic cytokinin hormone.

The most recently released mepiquat product is Stance, which consists of an 8.4 percent formulation of Mepiquat chloride, but also includes an additional active ingredient cyclanilide. Cyclanilide is an auxin transport and synthesis inhibitor. Stance consists of two products, each potentially regulating a different hormone involved in vegetative growth control.

### Vegetative Growth Management

Consider many factors when assessing the occurrence and potential for excessive vegetative growth including, growth stage, growth rate, fruit retention, availability of irrigation, field fertility status, field growth history, and previous history of applications of mepiquat chloride. No two fields or situations are identical and there is no universally accepted method of growth management. Not every field of cotton needs the same amount of mepiquat at the same time. There are several characteristics of the cotton plant which indicate the onset of excessive vegetative growth and should be considered when making mepiquat applications.

Regardless of the type of mepiquat product utilised or timing of application, the most effective vegetative growth management factor in cotton is high early-season fruit retention. The retention of early fruit reduces the amount of carbohydrates available for vegetative growth later in the season. Prior to any applications of mepiquat products, efforts should focus on maintaining an environment that promotes early fruit set and retention.

The growth stage of the cotton plant does much to determine the need for vegetative growth regulation. Prior to the development of the first square, or approximately 7 to 9 main-stem nodes, a cotton plant is strictly vegetative. Restriction of growth at this time has not proven necessary or effective. Once cotton has reached 5 nodes above white flower (NAWF) future vegetative growth is of little concern. Vegetative growth management can be limited to the time between the appearance of the first square and "cut-out" (5 NAWF), approximately 6 to 8 weeks during the growing season.

The height to node ratio (HNR) should be monitored to evaluate vegetative growth. The HNR is calculated by dividing the plant height by the number of main stem nodes, providing an average internode length. A cotton plant is considered to be vegetative when the average internode length is greater than a specific value at a specific developmental stage (Table 2).

*Table 2. Height to Node Ratios for normal, stressed, and vegetative plants.*

| *Crop Stage* | *Normal* | *Stressed* | *Vegetative* |
|---|---|---|---|
| | | HNR (inches/node) | |
| Seedling Cotton | 0.5 - 0.75 | – | – |
| Early Squaring | 0.75 - 1.2 | 0.7 | >1.3 |
| Large Square-1st bloom | 1.2 - 1.7 | <1.2 | >1.9 |
| Early Bloom | 1.7 - 2.0 | <1.6 | >2.5 |
| Early Bloom + 2 weeks | 2.0 - 2.2 | <1.8 | >2.5 |

This measurement alone can be deceiving however, in that all growth is included. Mepiquat chloride has no effect on growth that has already occurred; it will only affect growing plant cells.

The internode lengths of the uppermost five nodes on the cotton plant should also be monitored. Any internode located below this region has ceased expanding. The uppermost 5 nodes is the region of the plant actively growing which can be managed by applications of mepiquat chloride. Several tools have been developed to monitor the growth of cotton plants in this region. They provide guidelines for mepiquat chloride treatment. The Pix Stick, developed by BASF, provides a quick check of average internode length of the uppermost 5 internodes, and then gives a recommended Mepiquat chloride application rate.

Field history is also an important part of predicting and managing excessive vegetative growth. If the field has a previous history of rank growth, more attention should be paid to the use of mepiquat products for vegetative management. Conversely, fields prone to drought stress may require less or no use of mepiquat chloride.

While nitrogen (N) management is key to producing a cotton crop, low rates of nitrogen will lead to less vegetative growth and can also reduce lint yield. High rates of nitrogen can lead to rank growth, and potentially boll rot and delayed maturity. Cotton grown under high nitrogen fertility situations will require more mepiquat chloride to attain adequate vegetative growth control. The key is to target nitrogen rates so that yield is maximised while minimising excessive vegetative growth.

#### *Variety of Cotton*

Not all varieties of cotton exhibit the same growth habits,

which further confounds the vegetative growth management issue. For example, some cotton varieties have been bred for increased vigor that, while beneficial under cool early-season conditions, may be linked to excessive vegetative growth. Some full-season varieties tend to be more prone to excessive vegetative growth than shorter season varieties. Understanding the growth habits of a particular variety can be very useful for successful vegetative growth management.

Documentation shows not all varieties respond to mepiquat the same. Some of the new FiberMax offerings appear to be very responsive to mepiquat and may require a lower application rate and fewer overall treatments. Other varieties such as DP 555 BG/RR and ST 5599 BR require a very timely early season application in order to attain satisfactory vegetative growth control. Both seed companies and University of Georgia extension personnel evaluate the response of varieties to mepiquat applications. Especially with new varieties, it is important to pay particular attention to university and company-generated information in regards to mepiquat usage until some experience can be gained.

### Mepiquat Chloride Applications

Opinions about mepiquat chloride application rates, timing, and products differ greatly. General "progarmmes" can consist of a few high rate, multiple low rate, or plant growth based applications. All inclusive recommendations for mepiquat chloride applications are nearly impossible to make as excessive vegetative growth is influenced by many factors. The use rates of all materials containing a 4.2 percent concentration of Mepiquat chloride or a 9.6 percent concentration of mepiquat pentaborate are identical. Current research and grower experience indicates no difference in performance among these products regardless of additives or salt formulation (Table 3).

*Table 3. Comparison of height control and yield from three formulations of mepiquat vs. untreated check in small plot trials.*

| *Product* | *Height at Mid-bloom* | *Yield* |
|---|---|---|
| | *Inches/plant* | *Seed cotton/A* |
| Mepiquat chloride | 38.4 | 2991 |
| Mepiquat chloride + kinetin | 38.1 | 3090 |
| Mepiquat pentaborate | 35.7 | 2917 |
| Untreated | 50.7 | 2983 |
| Pr>f | 0.0001 | 0.8113 |
| C.V. | 9.1 | 12.0 |

## Growth Regulator Programmes

### *Irrigated cotton*

On irrigated cotton, two growth regulator progarmmes are recommended. One approach is to begin low rate applications of 4 oz/A during the second week of squaring. Applications should continue every 14 days for 3 or 4 applications. This approach was more applicable when producers made multiple trips across the field for insect control and mepiquat products could be added into the tank. With the adoption of Bt technology, producers may not make over the top applications of insecticides until late into the bloom period. Some varieties may require higher rates.

A more applicable approach is to treat with a higher rate (8 to 12 oz/A) at first bloom and subsequent higher rate treatments 2 to 3 weeks later if good growing conditions persist. For many cotton varieties, waiting to make initial applications until first bloom is an acceptable progarmme. However, with DP 555 BG/RR and ST 5599 BR in particular, waiting until first bloom to make the initial mepiquat application is often too late. Various research trials and grower experience has indicated that these varieties need to be treated during the squaring period.

*Non-irrigated cotton*

Applying mepiquat chloride to non-irrigated cotton is a more delicate task and plant monitoring is vital. While relatively uncommon, it is possible to apply too much mepiquat chloride if the treatment is followed by extremely dry conditions. If this occurs, cotton may be stunted and not recover even more favourable conditions return. Non-irrigated cotton can safely be treated with 8 oz/A at first bloom. Subsequent applications are then dependent upon adequate rainfall and growth, with rates and timings refined by monitoring overall plant growth. With ample early season rainfall many varieties should be treated prior to first bloom.

Calendar or growth-stage based progarmmes have weaknesses. The most successful progarmmes are those that consider the growth of plants on a field by field basis.

***Late-Season Mepiquat Chloride Applications***

Applications of mepiquat chloride are not recommended once cotton has reached 5 nodes above white flower. The theory behind such applications is that they would halt any further vegetative growth and force the plant to channel all growth into filling out top bolls.

### *PGR IV and Maxxon*

Cotton producers have shown steady interest in PGR IV as a plant growth regulator since its introduction in 1991. Manufactured by Micro Flo Company, it contains gibberellic acid and synthetic indolebutyric acid. The material has shown consistent growth and yield responses in some university trials. The label allows for treatments in the seed furrow, during the 1-7 leaf stage, at match head square and at early bloom.

Maxxon is a plant growth regulator similar to PGR IV, but it contains different quantities of gibberellic acid and synthetic indolebutyric acid. Manufactured by Riverside, it is labelled to be applied on cotton at match head square and first bloom. University research on Maxxon has been limited.

The effects of PGR IV and Maxxon on cotton has been studied at the MU Delta Center near Portageville since 1992. Both plant growth regulators were applied at 4 ounces at match head square and 4 ounces at early bloom. Response of cotton to both PGR IV and Maxxon ranged from 10 to 15 percent in 1992 to no differences in 1993. Though not statistically significant in these studies, both showed potential for higher lint yields.

The optimum timing for PGR IV on cotton is at planting in the furrow, at the second true-leaf stage and at the match head square stage. These applications had the greatest influence on cotton growth, leaf area and square retention. The manufacturer also promotes the material as beneficial with an early bloom application as well. Success may be made with this application if growth is not excessive. The optimum rate was 1 ounce per acre at planting and 4 ounces applied in a split between the second true leaf and match head square stages. To obtain maximum benefit from PGR IV or

Maxxon, use only when the crop is healthy and not stressed by heat, drought, insects or poor fertility.

Most research on the material has reported earliness, increased leaf area, increased nutrient uptake and square and boll retention as factors leading to higher yields. In-furrow treatments have shown slightly earlier emergence and enhanced seedling vigor. Researchers at the University of Texas have noted a tendency for more bolls to develop on the first two fruiting positions and on the lower 10 reproductive branches following treatment. Research has shown that at least two-thirds of the total lint produced comes from the first two fruiting positions. PGR IV or Maxxon may be options for growers who want to pursue high yields through a more intensive management approach.

Continued research will give us a better handle on the use of these materials. Until then, many farmers may want to try a conservative approach with plant growth regulator applications. It would be advisable to apply the materials early in the growing season to promote early growth, such as in-furrow (PGR IV) or match head square treatments, then use Pix or Mepichlor after bloom to slow vegetative growth and force a timely cutout.

Until further research is conducted, growers are advised not to tank-mix with Pix or Mepichlor. Since the active ingredient in these products inhibits gibberellic acid synthesis, their use in combination with PGR IV or Maxxon in a tank mix results in parts of the plant growth regulators working against each other. One promotes vegetative growth while the other slows it.

#### *Cytokin*

Cytokin was recently released by Plant and Bioregulator Technologies, Inc. for cotton production. The primary component of Cytokin is cytokinin. The product has shown some favourable results in university trials across the Cotton Belt, particularly when used in combination with Pix. University of Missouri research in southeastern Missouri has shown a yield advantage over the control treatment with the use of Cytokin applied at match head square and at early bloom. When applied in a tank mix with Pix, a greater lint increase was observed. This synergism was not noted with the combination of Pix and PGR IV or Maxxon.

Though university research on Cytokin is limited, the manufacturer recommends applying 2 to 4 ounces per acre beginning at match head square and applying weekly for a total of four applications. The hormone cytokinin is produced in the roots of plants. Under periods of plant stress, especially stress affecting the roots, cytokinin is likely to be inhibited. Cytokin is believed to replenish the plant's cytokinin supply during these periods.

Unlike the other plant growth regulators mentioned, Cytokin may be most beneficial under stressed conditions to maintain proper boll set and retention, promote earliness and develop a better root system. Cytokin or combinations of Cytokin and Pix have the potential for higher cotton yields, though growers are encouraged to use most plant growth regulators judiciously and on a limited basis at first, testing them under their own field conditions.

# 6

# Plant Growth Regulators Used in Floriculture

Plant growth regulators are used on a wide range of floricultural crops. Products currently on the market are labelled for control of plant height, for stimulation of lateral branching (Florel), for substituting for a cold storage requirement (GibGro 4LS), or for promoting flower initiation or earlier flowering. By far, the majority of plant growth regulators employed in floriculture are chemical growth retardants used for control of plant height. However, the application concepts and techniques are given below for all PGR's labelled for use on floricultural crops.

While digesting the following text, keep in mind that chemical growth retardants (CGR's) are a class of plant growth regulators (PGR's); not all PGR's are CGR's. Before applying a PGR, a grower should consider the reason for using the PGR. Recall that PGR's are applied to plants to regulate plant development and to stimulate a desired growth response. For example, chemical growth retardants are used to retard growth, resulting in shorter plants.

Most of the available growth retardants are anti-gibberellins; i.e., they inhibit the synthesis of gibberellins such as gibberellic acid (GA3) within the plant. Gibberellins stimulate cellular elongation, so without them, cells do not elongate as much, and plants do not grow as tall. Ethephon is not an anti-gibberellin; ethephon releases ethylene, which reduces elongation in some crops. Since CGR's, as do all PGR's, affect a specific process in the plant, it is essential they be applied in a manner that assures the most efficient response.

Both monetary and environmental costs are too great to apply PGR's carelessly. This conscientious effort to use the minimum amount of chemical as effectively as possible is the basis for Best Management Practices. Best Management Practices for using PGR's can be divided into different categories: timing, target tissue, dosage, application technique, and environmental conditions.

*Timing*: Timing of PGR applications must be matched to the proper plant stage of development to achieve the desired goal. Usually the labels on the products will give good descriptive stages of plant development to assist with correct application timing. For example, the GibGro label defines when to make applications. If GibGro 4LS is being used to partially substitute for cold storage of azaleas, plants should be at Stage 5 of floral development when treatment is initiated. Applying prior to this stage could result in abnormal flower development.

For most plants, apply CGR's such as A-Rest, B-Nine, Bonzi, Cycocel, Florel, and Sumagic after the plant has developed sufficient foliage to prevent excessive stunting of the plant's development. For example, a Bonzi spray on pansy plugs should not be made until three true leaves are present on the plants. Earlier applications may stop plant growth completely rather than merely reduce elongation. Apply CGR's just prior to rapid shoot elongation; e.g. after pinching and newly developing shoots are visible, but before shoot elongation has occurred.

Remember that CGR's are chemical growth retardants not chemical plant reducers-they cannot shrink plant growth already present. Make your final application before the stage when floral size will be reduced. If growth retardants are applied too late, the size of the flowers can be reduced and floral development can be slowed. A good example would be bract size reduction and delay of bract colour in poinsettia due to late application of A-Rest, B-Nine + Cycocel, Bonzi, Cycocel, or Sumagic. Timing of a PGR application should be based on a physiological stage of plant development such as the number of leaves present, the length of shoots, or plant diameter, not chronological age such as 3 weeks after pinching.

Recommendations given in terms of chronological age are merely guidelines that have been correlated to physiological stages and should only be used as rough estimates as to when to apply plant growth regulators. Also remember to read the label when deciding on timing for a PGR application. Many labels suggest the proper plant stage of development for chemical application.

*Target*: The target tissue or plant part to receive the chemical depends on the chemical and the plant species.

*Foliage and Stems*: Chemicals can be sprayed on, or shoots can be dipped into a chemical solution at time of transplanting. If dipping shoots into a solution, use a consistent soaking time, such as 10 seconds. For sprays, remember that areas are sprayed not individual plants. Imagine yourself painting a wall with a roller. You want uniform coverage over the entire area--not heavy clumps (every time you pass over a plant). This is the only way to assure even coverage and consistent results.

With many PGR's such as A-Rest, Bonzi, and Sumagic, spraying different volumes of the same concentration spray over a given area gives different levels of control. If growers "spray to runoff" or "spray to glistening", every grower will apply a different amount, and there is no way of predicting the results. Proper coverage is essential for consistent results. For example, Bonzi and Sumagic are not very effective if only applied to plant leaves.

Transport of Bonzi and Sumagic to the growing point in the plant is most effective if it is applied to the stem or if it is absorbed through the roots. If an insufficient application volume is used and stems are not wetted, poor height control will be the result. On the other hand, since root uptake is very effective, an over application of Bonzi or Sumagic using too large a volume will result in the excess spray dripping onto the substrate leading to root uptake of the chemical and causing too great a reduction in growth.

*Substrate*: PGR's can be applied to the substrate (growth media or soil), if the root system is the target tissue. This is the same as a soil drench; drenches use larger volumes of solution per plant or pot, but usually at lower concentrations than a spray or dip. Drenches can take more time to apply than sprays, and require exact metering of volume delivered per pot for consistent control of elongation.

*Roots*: PGR's can be applied directly to the roots or the underground portion of the plant such as bulbs, corms, or tubers, prior to planting. An example is a Bonzi soak, which is labelled as a preplant treatment for freesias.

*Dosage*: Read the label; do not guess on dosage. Keep in mind that a dosage is the product of concentration of solution applied and volume of solution applied per area. If either are incorrect, results could be unpredictable and nonrepeatable. Take care to always apply the correct dosage to all plants treated.

*Application Technique*: As with the target tissue, the method of delivering a PGR depends on both the chemical used and the plant species to be treated.

*Dips*: With some plants it is possible to dip the plant shoot, or underground portion (bulb, corm, tuber) into a growth retardant solution prior to potting. This method is labelled and is very effective for applying B-Nine to the shoots of rooted chrysanthemums just prior to potting. This method is fairly accurate, if each plant remains in the solution for the same amount of time, and if each plant has approximately the same size shoot or bulb. Unfortunately, this method is not feasible with many crops, and dipping plants in a common solution could result in the spread of disease organisms.

*Drenches*: Applying a growth retardant in a drench form is fairly easy. Measure out a known amount of chemical, add it to a known volume of water, and apply a known volume of the drench to each pot or plant. As pot size increases, usually the volume of drench recommended also increases. If using a drench, make sure the volume and concentration of the solution are correct for the size pot to be treated. There are "drenching machines" available on the market that emit a set volume of drench then turn off for a short period of time prior to the next dose to allow applicators to place the nozzle into the next pot to receive the drench. Do not apply any plant growth regulator through an irrigation system, unless the label lists chemigation as a legal application technique.

Currently, it is legal to apply A-Rest through drip, overhead boom, sprinkler, and flood (subirrigation) systems.

*Sprays*: A spray application can be more difficult to apply evenly than a drench, but with attention to detail, is the method of choice. Some chemical labels recommend to "spray to runoff"; that is, spray each plant until spray visibly just begins to drip off of the foliage. Depending on the size of the plant, the sprayer's objectivity, and other factors, varying amounts of chemical will be applied to each plant. It is much safer and more accurate to base spray application on areas, not plants. Apply a known volume of spray evenly to a known area (square footage), regardless of how many plants are in that known area.

The general recommendation for sprays is to apply ½ gallon per 100 $ft^2$ of cropping area. If the area is sprayed evenly, it assures that each pot will receive the same amount of spray, regardless of how many containers are in the area. To help with uniform distribution, the sprayer should be equipped with a pressure gauge and pressure regulator to assure uniform output. Unless the sprayer is emitting a constant volume, spray will not be applied evenly over plants. Spray droplet size can affect spray distribution on the plants.

In general, the smaller the droplet size, the more even the coverage and the greater the effectiveness (efficacy) of the chemical. Therefore it is possible that two growers applying the same treatment but with different spray nozzles could achieve different degrees of effectiveness. For consistency, try to use the same nozzle type for all PGR applications. Spray adjuvants, compounds added to assist the action of the active ingredient, such as a wetting agent can affect PGR efficacy. Some PGR products such as A-Rest, Bonzi, and Sumagic advise against adding a wetting agent.

B-Nine already contains a wetting agent and may not require the use of an additional wetting agent. If a wetting agent is needed to allow a spray material to more thoroughly cover plant surfaces, such as with a Cycocel spray on poinsettias, use as little as possible to avoid phytotoxic side effects. Start off using a very low concentration and by experimenting with water on test plants, slowly increase the wetting agent concentration. Add only enough wetting agent to prevent droplets from beading up on the leaves; a higher concentration is unnecessary and could cause leaf burning.

*Environmental Conditions*: The efficacy of a PGR can be affected by

- the environment at time of application,
- the status of the plant at application, and
- post-application treatment of the plant.

A good example of an environmental factor affecting PGR efficacy is the effect substrate components can have on CGR's. A barkcontaining substrate will reduce the efficacy of A-Rest and triazole growth retardants like Bonzi and Sumagic, when applied as a drench. Therefore, drenching may not be the application method of choice for these PGR's if using a barkbased substrate; or growers should account for the presence of bark when deciding on the concentration of PGR to use. Another environmental factor of concern is the time of day selected for applying PGR's. Research with foliar sprays of nutrients has shown that the time of day chosen for an application can affect a plant's ability to absorb a chemical.

Morning applications, when evaporation rates are low, are more desirable than later in the day. The longer the solution wets the tissue, the greater the chance for chemical uptake. If possible, make spray applications on a cloudy day to allow the chemical solution even more time to be absorbed prior to evaporation of the water from the surface of leaves and stems. Plant water status can affect chemical efficacy. Plants should not be wilted or stressed at application. A turgid plant is more able to absorb and translocate a PGR than a wilted plant.

With respect to post-application handling of treated plants, B-Nine has been shown to require a long period of time (up to 4 hours) for complete absorption after a spray application. If the plants are watered (wetting the foliage) too soon after a B-Nine application, the unabsorbed chemical will be washed off and the efficacy of the B-Nine will be reduced. This is not true for rapidly absorbed PGR's such as A-Rest, Bonzi, and Sumagic, and plants either drenched or sprayed with these chemicals can be watered as soon as 1 hour after application without a reduction in chemical efficacy. The effects of post-application watering on Florel and GibGro efficacy are not known.

Best Management Practices for plant growth regulators do not end with the guidelines given above. For a total programme, growers should always monitor the effectiveness of applied treatments to assure that the treatments are working and to help "fine-tune" the amount of chemical needed. Monitoring treatment effectiveness may indicate that less chemical is needed. Always leave a few untreated controls mixed in with the treated population. This allows comparisons for effectiveness of the PGR treatment.

For growth retardants, monitor plant growth for the treated plants regularly to know when plants begin to "grow out" of the treatment and to help in deciding if and when another application is needed. When used properly, chemical plant growth regulators are effective tools to help produce a high quality crop. When misused, they can reduce crop quality and increase production costs. Uniformity and consistency in application are crucial to attain predictable and desirable results.

Benzyladenine at 100 ppm is highly effective on holiday cactus in stimulating branching when applied during the vegetative phase and is effective in increasing the number of flower buds when applied during reproductive conditions; but no product is labelled for this use. Accel (N-Benzyl-9-[2-tetrahydropyranyl]adenine [BPA] is the active ingredient) was labelled for increasing lateral branching in carnation and roses, but will not be available after current supplies are depleted. Research on floricultural crops has shown many other potential uses for PGR's, such as gibberellic acid ($GA_3$) substituting for cold storage of hydrangea and hastening flowering of cyclamen; however, only those uses listed on a product label can be implemented legally.

Plant growth regulators are regarded as pesticides, and it is a violation of Federal and State Law to use these products in a manner inconsistent with their labelling. Hopefully, expansion of current labelling will be possible in the future to allow growers to take advantage of research results showing more efficient cropping and higher quality plants through best management utilisation of plant growth regulators.

# 7

# Tree Growth Retardants

In the past, costly mechanical trimming was the sole method available to arborists and utility foresters to reduce tree and shrub size. Chemical growth retardants were developed as an inexpensive approach to limit size and the growth rate of trees and, at the same time, to enhance their tolerance to the harsh environmental conditions of urban areas.

Utility arborists were the first among those caring for trees to peer over the fence at agricultural and horticultural fields and ponder the potential of growth regulators used in those cropping systems as a tool for tree maintenance. Mechanical trimming, which was the sole means to combat the unrelenting growth of trees into overhead electrical wires, was a costly operation and a chemical alternative was very attractive. Hence, the electric utility industry provided funding in the late 1950s for research on chemical control of tree growth following trimming for electric line clearance. Results of that early research led to the use of napthaleneacetic acid (NAA), a synthetic auxin, painted onto the surface of pruning wounds.

Although effective in reducing the regrowth of branches, coating each cut surface high in the crown of trees took a lot of time and was not cost effective. Hence, in the 1970s new TGRs and more economical application techniques were sought. The first major breakthrough in the commercial feasibility of TGRs on a large scale was the formulation in the late 1970s of the cell elongation inhibitors, paclobutrazol, uniconazole, and flurprimidol for trunk injection. Due to their low water solubility, it was considered necessary to dissolve the new generation of growth retardants in either methyl or isopropyl alcohol.

The active ingredients of these formulations were unquestionably effective in reducing tree growth. After several years of use throughout the United States in the 1980s, problems associated with trunk injection begin to appear. Cracks in the bark and cambium, weeping from injection holes, and internal wood discoloration due to the alcohol carriers led to disenchanted utility arborists and their customers. A decline in use of TGRs followed. Uniconazole was even removed from the tree care market. However, in spite of these problems, interest among utility arborists continued in a chemical tool to reduce trimming frequency and the amount of wood waste removed from trees.

Flurprimidol, sold as Cutless Tree Implants®, was pressed into tablets for insertion into shallow holes drilled in tree trunks. Concern about drilling holes into trees and the apparent compartmentalisation around the tablets that prevented continued slow release of flurprimidol into the transpiration stream resulted in limited use of the implants. Hence, flurprimidol was removed from the tool kit of arborists about two years ago.

Today, only one growth retardant for use on trees remains, paclobutrazol. Satisfactory performance of paclobutrazol as a growth retardant, as well as several benefits to tree health, revealed through recent research that resulted in a rebound in use of this TGR today by some electric utilities and spurred an active expansion of the market to commercial landscapes and general arboricultural tree care.

Paclobutrazol, formulated as Cambistat 2SC® or Profile 2SC®, is applied as a water suspension. Both formulations are approved by the EPA for soil injection or application as a basal drench. The dose rate, which is species specific, is determined by measuring trunk diameter. The water suspension of paclobutrazol can either be injected at about 150 psi into the soil to a depth of approximately 6 inches as close to the tree trunk as possible or simply poured into a shallow trench around the base of each tree. The product label provides detailed information for proper application.

Treatments can be made anytime the soil is not frozen or saturated with water. Actually paclobutrazol and other growth retardants with the same mode of action are currently used in the nursery industry for production of compact and hardy bedding plants and on golf courses to reduce growth of turf and the frequency of mowing fairways. The dose rate for turf is lower than that applied to trees.

Consequently, the grass in a narrow ring around the base of paclobutrazol-treated trees may be notably shorter. However, this could be a benefit because the serious problem of mower and string trimmer damage to tree trucks is less likely without the need to mow close to trees. Since paclobutrazol in very immobile in soils, there is no need for concern about over-regulation of turf more than a few inches away from the treatment zone.

Suppression of growth by paclobutrazol occurs because the compound blocks three steps in the terpenoid pathway for the production of the hormone gibberellin by binding with and inhibiting the enzymes that catalyse the metabolic reactions. One of the main roles of gibberellins in trees is the stimulation of cell elongation. When gibberellin production is inhibited, cell division still occurs, but the new cells do not elongate. The result is shoots with the same numbers of leaves and internodes compressed into a shorter length.

For many years this was considered to be the sole response of trees to treatment with paclobutrazol. However, recent research has demonstrated that blocking a portion of the so-called terpenoid pathway causes shunting of the accumulated intermediary compounds above the blockage. The consequence is increased production of the hormone abscisic acid and the chlorophyll component phytol, both beneficial to tree growth and health. The unique structure of paclobutrazol that allows it to bind to an iron atom in the enzymes essential for the production of gibberellins also has the capacity to bind to enzymes necessary for the production of steroids in fungi as well as those that promote destruction of abscisic acid.

The consequence is that paclobutrazol treated trees have greater tolerance to environmental stresses and resistance to fungal diseases. Morphological modifications of leaves induced by treatment with paclobutrazol such as smaller stomatal pores, thicker leaves, and increased number and size of surface appendages on leaves may provide physical barriers to some fungal, bacterial, and insect infestations.

## Reduction of Growth

### *Shoot Growth*

Although growth reduction is dose sensitive and varies widely among species, all evergreen and hardwood species, and even palms, respond in some degree to treatment with paclobutrazol. Treated trees have more compact crowns and somewhat smaller and darker green leaves, but otherwise look normal. The amount of shoot growth reduction ranges from a low of 10 percent to a high of 90 percent, with average growth reduction being 40 to 60 percent when recommended dose rates are applied. As a consequence of the reduced growth in height, there is a parallel reduction in biomass removed when trees eventually require trimming.

### *Cambial Growth*

Although the principal focus of research with paclobutrazol has been on growth in length of shoots, reduced growth in diameter of the trunk and branches of woody plants also has been found. Expansion of cells produced by the vascular cambium also depends on gibberellins just like cells in stems and leaves. This could have significance in urban areas for trees planted in wells, above ground containers, and in the parkway between sidewalk and curb. Up to 30 percent of trees planted in the city cause sidewalk and curb damage due to expansion in girth of the trunk and roots, requiring significant portions of annual tree budgets for costly repairs. Suppression of diameter growth of tree trunks and roots at least forestalls costly damage and the creation of hazards.

### *Root Growth*

Effects of paclobutrazol on root growth vary from

enhancement to inhibition and are far from being clearly defined and understood. In almost all cases, however, the response in paclobutrazol-treated trees is an increase in root to shoot ratio. Gary Watson at the Morton Arboretum conducted one of the few studies on large mature trees exposed to *paclobutrazol*.

Soil injection at the base of white and pin oaks caused fine root densities to be 60 or 80 percent higher, respectively, near the trunk base. It is unclear whether the responses observed in roots of treated trees are a direct effect of paclobutrazol on root growth or an indirect effect resulting from shoot growth modification and a shift in carbohydrate allocation to the roots. Root response to paclobutrazol is an important question because root growth and vigor influence not only water uptake but many other aspects of tree health.

### Greener Leaves

Trees treated with paclobutrazol generally have leaves with a rich green colour suggesting higher chlorophyll content. There are two possible explanations for this response. One is that the leaves of both treated and untreated trees contain the same number of cells, but because the cells in leaves of treated trees are smaller, the chlorophyll is more concentrated in the reduced cell volume.

In addition, however, there is evidence that the amount of chlorophyll is actually increased too because phytol, an essential part of the chlorophyll molecule is produced via the same terpenoid pathway as gibberellins. Paclobutrazol treatment, which blocks the production of gibberellins, results in a shunting of the intermediate compounds from gibberellin synthesis to the production of even more phytol.

*Figure 1. Sugar maple leaves from trees untreated or treated with paclobutrazol showing higher chlorophyll content.*

An analogy might be an accident blocking the flow of traffic on a major highway causing more drivers to divert to alternate routes.

### Reduced Water Stress

In addition to interfering with gibberellin production, paclobutrazol is known to affect the synthesis of the hormone abscisic acid. Abscisic acid also is made via the terpenoid pathway. Unlike the inhibiting effect on gibberellin synthesis, treatment with paclobutrazol promotes the production of abscisic acid much like it promotes the production of phytol. When gibberellin synthesis is inhibited, more precursors in the terpenoid pathway accumulate and are shunted to the production of abscisic acid. Paclobutrazol also interferes with the normal breakdown of abscisic acid.

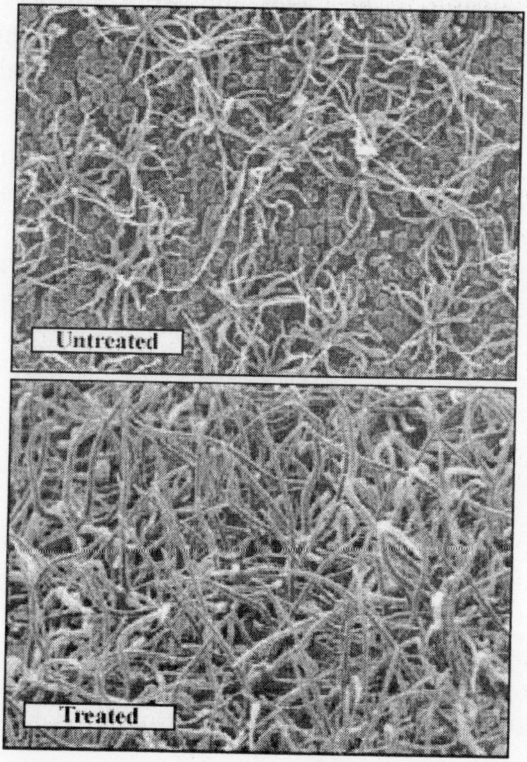


*Figure 2. Scanning electron micrographs of the lower surface of leaves of cherrybark oak untreated or treated by the soil injection method with paclobutrazol.*

The mode of action involves another iron containing enzyme to which the paclobutrazol will attach, preventing its activity. The combined effect on both the production and breakdown processes results in enhanced concentrations of abscisic acid in leaves. One of the functions of abscisic acid is to cause stomates to close, reducing water loss from leaves through transpiration. Improved water relations in trees could arise from a combination of increased abscisic acid contents that physiologically reduce stomatal opening, reduced shoot growth resulting in less leaf and stem surface area for transpiration, more fine roots to absorb water, and structural changes in leaves that provide physical barriers to moisture loss.

Fig. 2 shows dramatic scanning electron microscope images of thicker leaves and masses of hairs on leaf surfaces of cherrybark oaks in response to treatment with paclobutrazol. The improvement of water relations in paclobutrazol treated trees is an important secondary benefit of using a TGR.

### Protection from Fungal Diseases

Protection from fungal diseases that attack urban trees is now recognised as another secondary benefit of using paclobutrazol. There are numerous observations of reduced incidence of common fungal diseases such as anthracnose following treatment with paclobutrazol. Karel Jacobs at the Morton Arboretum has shown paclobutrazol to significantly reduce the growth of eight fungal pathogens in laboratory cultures.

The fungistatic property of paclobutrazol is due to the inhibition of steroid production in fungi, also via the terpenoid pathway. This is the same mode of action that accounts for the fungistatic property of the class of fungicides known as SBIs or steroid biosynthesis inhibitors. Steroids are essential constituents of membranes. The increased resistance of paclobutrazol-treated trees to bacteria is not thought to be a direct effect on the pathogen, but rather due to alteration in leaf surface structure or even the size of stomatal pores that make infection more difficult.

# 8

# Plant Growth Regulators on Containerised Herbaceous Perennials

A perennial plant or perennial is a plant that lives for more than two years. Herbaceous perennials are plants that do not form woody tissue and woody perennials are plants that develop a woody base or root system from which the foliage and flower stems grow. The term perennial more commonly describes herbaceous perennials, since woody plants (i.e., trees and shrubs) are always perennials. Perennials that flower and fruit only once and then die are termed monocarpic or semelparous. However, most perennials are polycarpic, flowering over many seasons in their lifetime.

In warmer and more clement climates, perennials grow continuously. In seasonal climates, their growth is limited to the growing season. For example, in temperate regions a perennial plant may grow and bloom during the warm part of the year, with the foliage dying back in the winter. These plants are deciduous perennials. Regrowth is from existing stem tissue.

In many parts of the world, seasonality is expressed as wet and dry periods rather than warm and cold periods. In some species, perennials retain their foliage all year round; these are evergreen perennials. With their roots protected below ground in the soil layer, perennial plants are notably tolerant of wildfire. They are also less subject to extremes of cold in temperate and arctic winters, with less sensitivity than trees or shrubs.

Perennial plants dominate many natural ecosystems on land and in fresh water, with only a very few (e.g. Zostera) occurring in shallow sea water. They are particularly dominant in conditions too fire-prone for trees and shrubs, e.g., most plants on prairies and steppes are perennials; they are also dominant on tundra too cold for tree growth. In forests, perennial plants are of secondary importance to trees and shrubs, but are often still abundant on the forest floor.

Perennial plants are usually better competitors than annual plants, especially under stable, resource-poor conditions. This is due to the development of larger root systems which can access water and soil nutrients deeper in the soil and to earlier emergence in the spring.

There is a tremendous diversity of herbaceous perennial plant species being grown for both the retail and landscaping sectors of the industry. Because of the diversity in species grown, there is much more unknown about perennials production than is known. Growth regulation is of particular concern. In production settings, as well as in retail locations, herbaceous perennials grown in pots tend to stretch and become leggy or simply overgrow their pots before their scheduled market date. These plants are less marketable, and harder to maintain.

Many growers resort to pruning, which is not only costly in terms of labor, but also delays plant production two to four weeks. Availability of chemical plant growth regulators (PGRs) for perennials is not a problem. Due to recent label expansions, all of the primary floriculture growth retardants are labelled for use on perennials. However, lack of knowledge about rates and the diversity of plant responses to these PGRs are problems. Many of the herbaceous perennials in the market have never been evaluated for response to any of these chemicals. Foliar spray applications are tested more often than drenches, presumably due to the higher labor costs involved in applying drenches in large scale production areas.

Pay particular attention to application instructions on PGR labels. For soil active PGRs like A-Rest, Bonzi and Sumagic, the dose applied to the plant is a function of both the solution rate (ppm active ingredient) and the volume applied to plant and its substrate. For foliar sprays, all plants should be treated according to a specified volume of PGR applied to a specified area of bench, not to the individual plants. Generally, the labels recommend 2 qt. per 100 sq.ft. of bench which is sufficient to cover the plant and permit a small amount of runoff onto the medium.

While 50 ppm Bonzi applied at the label recommended volume resulted in acceptable height control of hollyhock (Alcea rosea), application of the same rate at twice the recommended volume resulted in excessive growth reductions (Figure 1). Treatment of large or dense plants may require higher volumes to treat the interior portions of the plant. However, remember that deviations from the recommended volumes applied will result in a different dosage of PGR applied.

Plant growth regulators should be applied to new growth (1.5 to 2" long) before it enters the rapid elongation phase. Generally, PGRs should be applied before flower initiation to minimise flowering delays. Remember that reductions in plant height are not always the growth control desired during production.

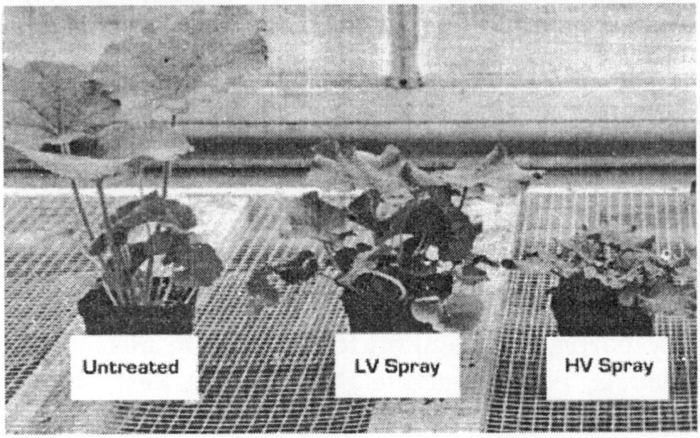


*Figure 1. Application volume affects plant response to soil-active PGRs like Bonzi: hollyhock (Alcea rosea) treated with 50 ppm Bonzi: untreated control (left), at label recommended volume of 2 qt. per 100 sq. ft. (center) or at twice the label recommended volume (4 qt. per 100 sq. ft.).*

For gaura (Gaura lindheimeri), where excessive height was not a production issue, B-Nine or a B-Nine/Cycocel Tank Mix was very effective in reducing plant width (Figure 2).

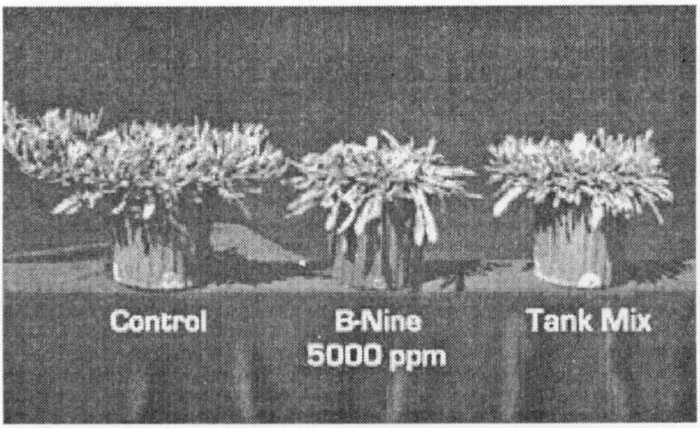


*Figure 2. PGRs may affect plant width more than height: gaura (Gaura lindheimeri 'Siskiyou Pink') untreated (left), 5000 ppm B-Nine applied twice (center), and a Tank Mix of 5000 ppm B-Nine plus 1500 ppm Cycocel applied once (right).*

## Important Plant Growth Regulators

B-Nine (daminozide, Uniroyal Chemical Company) is the most commonly used PGR in the floriculture industry. In general, it is not phytotoxic, and has a short-term effect that seldom results in overstunting of treated plants. Due to the low activity of B-Nine and its lack of soil activity, it is easier to apply consistently than the newer, more potent PGR chemistries. The low activity also means that B-Nine must be applied more frequently to maintain control over vigorous crops.

Nearly one-half of the perennials tested have shown some response to multiple applications of B-Nine. Generally, foliar sprays of 5000 ppm are applied every 10 to 14 days as necessary. B-Nine is labelled for use on containerised or bed-grown crops in the greenhouse, and on containerised plants grown outdoors under nursery conditions. Frequency of application may need to be increased to weekly for more vigorous cultivars grown outdoors.

Cycocel (chlormequat chloride, Olympic Horticultural Products) is another PGR with a long history in floriculture. Cycocel is generally applied as a foliar spray at 1500 to 3000 ppm. Although rates above 1500 ppm often cause chlorosis on treated leaves of other floricultural crops. However, Cycocel alone has not been tested on a wide variety of perennials. It was very effective on Campanula carpatica and purple coneflower (Echinacea purpurea) where multiple applications of 1500 ppm resulted in excessive reductions in growth.

Rose mallow (*Hibiscus moscheutos*) also is sensitive to Cycocel, responding well to multiple applications of 1000 ppm. Cycocel also promotes earlier flowering and greater flower numbers on *Hibiscus*. First application should be made when the laterals are 0.5 to 1 inch long.

A B-Nine/Cycocel Tank Mix has more PGR activity than either B-Nine or Cycocel alone and has been tested on a wide variety of perennials. Three-lobed coneflower (*Rudbeckia triloba*) was very responsive to B-Nine applied twice at 5000 ppm, but not responsive to Cycocel at rates up to 4000 ppm. A tank mix of 5000 ppm B-Nine with increasing rates of Cycocel resulted in similar height control with a single application at four weeks prior to the photograph.

The "high activity" rate generally used for the tank mix is 5000 ppm B-Nine plus 1500 ppm Cycocel. Although the rate of B-Nine is usually adjusted to increase or decrease activity, changing the Cycocel rate also affects activity. As described with speedwell, single applications of the tank mix may be more effective than multiple applications of B-Nine alone.

A-Rest (ancymidol, SePRO Corporation) is a more active compound than B-Nine or Cycocel. It is labelled for use on containerised plants in greenhouses and nurseries. A-Rest is active as a spray or a drench so application volume affects plant response. In addition, A-Rest is one of the few PGRs labelled for distribution through the irrigation system via flood, sprinkler or drip systems. Although it has not been evaluated for effectiveness on a large number of perennial species, A-Rest is specifically labelled for foliar sprays on bleeding heart (*Dicentra*) and columbine (*Aquilegia*) at 65 to 132 ppm, gayfeather (*Liatris*) (25 to132 ppm) and hybrid bee delphinium (*Delphinium*) (35 to 132 ppm). A-Rest also can be applied as a drench at 2 to 4 ppm for these crops.

For best results, treatments should be applied to well-rooted plants in active growth, prior to the initiation of flowering. In research reports, tickseed (*Coreopsis*) was not responsive to A-Rest, but Campanula, hardy ageratum

(*Eupatorium*), purple coneflower (*Echinacea*), lavender (*Lavandula*), cardinal flower (*Lobelia*) and speedwell (*Veronica*) were very responsive. The higher rates necessary for foliar applications to some crops may not be economically feasible. The use of A-Rest as a drench or as a treatment of plants in the plug stage is more economical than foliar application on finished plants.

Bonzi (paclobutrazol, Uniroyal Chemical Company) and Sumagic (uniconazole, Valent USA) are members of the triazole class of PGRs and are much more active than the previous compounds. Sumagic is more potent than Bonzi. For perennials, Bonzi is generally applied at rates of 30 to 100 ppm and Sumagic at 15 to 45 ppm. These PGRs are rapidly absorbed by plant stems and petioles or through the roots. Excess spray dripping off treated plants acts as a drench to the substrate, increasing the activity of the treatment. For foliar sprays, uniform application of a consistent volume per unit area is critical to uniform and consistent crop responses to the triazoles.

Both compounds are labelled for application to the media surface prior to planting plugs. In this case the PGR is applied as a spray (at rates one-third to one-half the recommended foliar spray rate) to the surface of the medium in filled pots. The PGR moves into the medium with subsequent irrigations and effectively behaves as a drench which is the reason for the lower recommended rate. Effectiveness also is reduced by bark in the medium as it is with drenches.

Neither PGR has exhibited any specific phyotoxicity symptoms on perennials, but care must be taken with application of excessive rates on sensitive plants. In some cases, excessive stunting can be very persistent, e.g., goldenrod (*Solidago sphacelata*) treated with higher rates of Bonzi (240 ppm) or Sumagic (60 ppm) did not recover normal size at five months after planting into the landscape. Growth of velvet sage (Salvia leucantha) was excessively reduced by 45 or 60 ppm Sumagic in the greenhouse (Figure 3a). Furthermore, 60 ppm Sumagic caused a significant delay in landscape growth (Figure 3b).

*Figure 3. Greenhouse applications can cause significant delays in the resumption of growth in the landscape: velvet sage (Salvia leucantha) a) in the greenhouse at five weeks after treatment with one application of Sumagic at 0, 15, 30, 45, or 60 ppm (left to right); b) at eight weeks after planting into the landscape, 12 weeks after treatment with 60 ppm Sumagic (right) or untreated (left).*

These compounds must be used carefully and appropriately. Especially when working with the triazoles, thoroughly test your application methods and rates on a small number of plants before treating your entire crop.

Avoid late applications of the triazoles. They should be applied prior to flower initiation when possible. The persistence of these compounds in plant stems and petioles can have significant effects on the flower display as well. As with many of the growth retardants, the triazoles inhibit gibberellin synthesis. Generally, the most rapidly elongating tissues have the highest production of gibberellins and, therefore, are most affected by reductions in gibberellin production. For example, elongation of the flower inflorescences of gaura (*Gaura lindheimeri*) was much more sensitive to growth inhibition than was the elongation of the stem tissues (Figure 4).

*Figure 4. Elongation of flower inflorescence may be more affected by PGRs than elongation of stem tissue, as seen with gaura (Gaura lindheimeri 'Siskiyou Pink') treated with increasing rates of Sumagic (0, 15, 30, 45 or 60 ppm, left to right).*

Usually this is acceptable because it keeps the flower height in better proportion to the plant height. However, differences in response vary, and in some yarrow (Achillea) cultivars, flower height has been excessively reduced at moderate application rates. Bonzi has a broad label for ornamentals that includes use on greenhouse or outdoor grown containerised crops.

Bonzi also is labelled for application through the irrigation system or by subirrigation, including ebb/flow or flooded floor systems. Bonzi has been tested on a wide variety of perennials with species ranging from extremely sensitive to low rates to non-responsive to very high rates. In many cases, multiple treatments with lower rates have been more effective, with less chance of overstunting, than a single application at a higher rate.

Sumagic also has a broad label for ornamentals, but its use is limited to containerised plants grown in greenhouses, overwintering structures, shade houses, or lath houses. It is not labelled for outdoor nursery use. At this time, Sumagic is not labelled for application through any irrigation system. Sumagic has been very effective on a large number of perennials. Test rates in the 15 to 45 ppm range. Since it is very potent, pay special attention to uniform application and proper volumes. Use caution in the higher rates or on more sensitive species since Sumagic can be very persistent in the landscape (Figure 3b).

Florel Brand Pistill (Florel) (ethephon, Monterey Chemical) is a compound that breaks down in plant tissue after application to release ethylene, a natural plant hormone. As with ethylene, its effects can vary depending upon the species and the stage of growth at time of application. It has a new broad use label for increasing lateral branching of floricultural crops. Florel also inhibits internode elongation of many plants; however, research with perennials has been limited.

Florel has controlled runner elongation of clump verbena (Verbena 'Homestead Purple'), and increased inflorescence numbers of sage (Salvia 'May Night') and yarrow (Achillea). Florel should be applied to actively growing plants prior to flower development. If flowers are present, they are likely to abort. Florel may delay flowering about one to two weeks, particularly if applied close to time of flower initiation. Florel should not be applied to plants that are heat or drought stressed. The pH of the water used for the spray solution can be important. If the pH is too high, the ethephon will convert to ethylene before it gets to the plant and activity will be reduced.

Florel contains sufficient acidifiers and buffers to maintain a pH of 5.0 or lower when mixed with most greenhouse water supplies. In general, water that has sufficient quality for irrigation of greenhouse crops (moderate pH and alkalinity) is suitable for mixing Florel. However, if you are acidifying your water prior to irrigation, use the acidified water for mixing the Florel as well. The solution should be applied within four hours of mixing.

Atrimmec (dikegulac sodium, PBI Gordon) is a compound that interferes with terminal growth by inhibiting deoxyribonucleic acid (DNA) synthesis which is required for new growth. By primarily inhibiting terminals, apical dominance is reduced which enhances the production of lateral branches. This mode of action tends to cause a delay in the resumption of plant growth that may add 2 to 4 weeks to production time. Atrimmec is labelled for use on containerised and landscape woody ornamentals, but also is labelled for use on some hanging basket plants and groundcovers.

Of particular interest to perennials growers is its label for lantana (*Lantana camara*) (750 to 1500 ppm) and butterfly bush (*Buddleia*) (530 to 1500 ppm). Atrimmec should be applied to actively growing plants with at least two nodes to provide sufficient lateral development. In addition to creating a more full plant, enhancing the number of laterals in a pot generally reduces the overall height of the plant due to the greater distribution of resources. Responses are very species specific so test several rates under your growing conditions. Atrimmec usually causes leaf chlorosis which can be very persistent at higher rates (above 1500 ppm). Other phytotoxic responses, including malformed flowers, have been noted at higher rates on perennials such as gayfeather (*Liatris spicata*).

# 9

# Commercial Greenhouse Flowers

Many greenhouse-grown floricultural crops tend to grow taller than desired and require height control measures to prevent excessive internodal elongation. There are three categories of control available to producers-biological, physical, and chemical control. Each should be given consideration by producers who often produce undesirably tall crops. When deciding on the best control method, consider the cost of the method (including equipment, labour, and other encountered expenses such as fuel); consider how the method will affect crop timing; and consider how the method will affect plant quality in the greenhouse, the retail outlet, and in its place of use.

Biological control is theoretically the ultimate method. Through genetics and plant breeding, a plant cultivar is developed that grows to the perfect height. Often there are cultivars which are genetically short. For example, marigold cultivars can range in height from 5 to 30 inches. Select cultivars that best match your production system and market demands. Unfortunately, there are not perfect height-cultivars for all of the greenhouse crops and growers must rely on other control mechanisms other than biological control of height.

Physical control of plant height is based on knowing how the growing environment and cultural practices affect plant growth habits. Physical methods of control may include:

- *Container size*: Size restriction of the root system can be used to slow the development/reduce the stretching of bedding plants.
- *Water stress*: Allowing plants to wilt slightly between waterings leads to shorter plants, but quality may be greatly reduced. In some instances, such as with impatiens seedlings, it can be used quite effectively without long-term damage to the plants. Care should be taken not to subject plants to a wilt so severe that they cannot recover, or that leaf or flower damage occurs.
- *Nutrient stress*: Withholding or reducing fertiliser tends to slow overall growth, including stem elongation. Reducing nitrogen fertilisation to plants is effective in height control for some plants such as impatiens, but also slows down the growth and development of the plants along with stem elongation, and can reduce plant quality. Withholding phosporus application is a common practice to keep plug seedlings short.
- *Wind, vibrations or touching*: Stem elongation is less when plants are exposed to wind, vibrations, or routine touching. These "seismomorphogenic" responses (responses to vibrations/wind) and "thigmomorphogenic" responses (responses to touching/brushing) can be used for height control, but are not economical in many cases and can spread foliar diseases. An exception would be with tomato and other vegetable transplants where there

are no chemical height controls available. For example, brushing tomato transplants (foliage should be dry) for about 40 strokes back and forth with a cardboard tube suspended from an irrigation boom twice daily for 18 days can result in as much as a 30% reduction in stem elongation. This technique also works for Easter lilies, but again, the economics and labour requirements may be prohibitive and should be evaluated prior to implementation.

– *Light intensity*: Higher light intensity or irradiance tends to reduce plant elongation, and plants will be shorter at maturity. The key wording is "shorter at maturity". Higher irradiance increases plant growth rate, and plants will grow faster at first than plants in lower irradiance. However, at maturity or flowering, the plants produced in higher light usually will be shorter. Given very poor light, plants tend to stretch, become leggy, and develop more slowly. Placing plants too close together reduces the amount of light each plant receives and will result in excessive stretching as well. Use correct spacing of plants for best growth and reduced elongation.

– *Photoperiod*: In general, short days of winter will result in both less growth and less height than long days of summer. The reduced growth is due to less light being received per day by the plants while the reduced elongation is a direct photoperiod response related to the daylength, not associated with lower irradiance levels. However, photoperiod is not a factor usually regulated strictly for height control since other effects such as flowering can also be affected by photoperiod.

- *Light quality*: Recent research has documented that the high far red to red light ratio found during twilight and dawn is responsible for increased plant height. By drawing blackcloth over Easter lilies prior to twilight (about 30 minutes prior to standard sunset) and leaving plants covered until full sunrise (about 30 minutes after standard sunrise), a 20% reduction in height can be achieved. Other crops are probably also affected by eliminating twilight and dawn light. Another aspect of light quality that can be utilised to prevent excessive plant height is filtration of the longer (far red) wavelengths. Certain blue filters and blue tinted plastic greenhouse coverings can be effective in reducing internode elongation by changing the quality of light (filtering out certain wavelengths) plants receive.
- *Temperature*: There are two main aspects of temperature that can affect plant growth and height. The average (average of day plus night) temperature affects plant growth rate; higher averages usually result in more rapid growth and development. The difference between the day and night temperature (DIF) affects stem elongation and plant height. The DIF is calculated as: day temperature (°F) minus night temperature (°F) and can be positive (day higher than night), zero (day = night), or negative (day less than night). Negative DIF's reduce stem elongation, compared to zero and positive DIF's and can be used to control plant height. Lowering the temperature for a 2 hour period starting 30 minutes prior to dawn is almost as effective as keeping a lower temperature throughout the entire day. This is fortunate for

growers in North Carolina, since temperatures during most of the year prohibit maintaining a 60°F day temperature for the entire day. DIF is effective on many crops including many bedding plants, poinsettias, mums, and Easter lilies.

A suggested DIF programme for many plants such as poinsettias and most bedding plants would be to use a 68°F night temperature, a 2 hour drop to 60°F beginning 30 minutes prior to dawn, and a 65°F day venting temperature setting. For warm-grown plants including begonia, vinca, and verbena, use a 72°F night temperature, a 64°F 2 hour drop beginning 30 minutes prior to dawn, and a 65°F day venting temperature setting. Be aware that night temperatures above 68°F can cause delay in floral initiation of some crops such as poinsettias; these night temperatures are best avoided, except when recommended for specific crops.

Also a delay in flowering may occur if average daily temperature (average of day + night temperature) is reduced below about 62°F; lower daily averages slow growth and development rates. Too great a negative DIF (greater than minus 10°F) should not be used; it can cause leaves to droop and appear wilted. Another concern with DIF is timing of the application with respect to flower development. Do not use DIF late in the cropping cycle or flower and bract size can be reduced. For example, DIF should be discontinued on a poinsettia crop prior to visible bract color.

Chemical control involves the use of chemical growth retardants. Most of the available growth retardants are antigibberellins, as they inhibit gibberellin (GA) synthesis within the plant. GA's are responsible for cellular elongation, so without them cells do not elongate as much, and plants do not grow as tall. Ethephon (Florel®)

is not an anti-gibberellin. Ethephon releases ethylene, which reduces elongation in some bulb crops. The chemical growth retardants labelled for use in floriculture differ in their basic characteristics and affect growers' choices of chemicals as well as determine when and how to apply.

*Timing*: For most plants, apply after the plant has developed sufficient foliage (photosynthetic area, leaf area) to prevent stunting of the plant's development. An example appears on the Sumagic® label, recommending a minimum height of 4 inches for pansies prior to a spray application. Apply just prior to rapid shoot elongation; e.g. after pinching and newly developing shoots are visible, but have not yet elongated. Make your final application before the stage when floral size will be reduced. If growth retardants are applied too late, the size of the flowers can be reduced and floral development can be slowed.

A good example would be bract size reduction in poinsettia due to late application of Cycocel® or Bonzi®. Remember that timing of application should be based on a physiological stage of plant development such as number of leaves, length of shoots, or plant diameter, not chronological age such as 3 weeks after pinching. Recommendations given in terms of chronological age are merely guidelines that have been correlated to physiological stages and should only be used as rough estimates as to when to apply growth retardants.

*Target*: The target tissue or plant part to receive the chemical depends on the chemical used and the plant species being grown.

– *Foliage and Stems*: Chemicals can be sprayed on, or shoots can be dipped into the solution at time of transplanting. If dipping shoots into a solution, use a consistent soaking time, say 10 seconds. For sprays, remember that areas are sprayed, not individual plants. This is the only way to assure even coverage and consistent results. If growers "spray to runoff" or "spray to glistening", every grower will apply a different amount, and there is no way of predicting the results.

– Substrate: Apply to the substrate (growth media or soil). This is the same as a soil drench; drenches use larger volumes of solution per plant/pot, but usually at lower concentrations than a spray or dip. Drenches can take more time to apply than sprays, and require exact metering of volume delivered per pot. There are machines available that automatically deliver exact volumes to automate drenching procedures.

– Roots: Apply directly to the roots/underground portion of the plant. This method is restricted to plants such as dahlia tubers that are potted by the grower. Growth retardants can be applied as a dip or soak in solution to roots, bulbs, corms, tubers, or crowns. An example is a Bonzi® dip, which is labelled as a preplant soak for amaryllis, caladiums, calla lily, daffodil, freesias, hybrid lily, montbretia, tulips, and other bulb species. One major problem with bulb soaks noted in research with Easter lilies has been inconsistent results and high plant-to-plant variability in response.

*Dosage*: Read the label; do not guess on dosage. Keep in mind that a dosage is the product of [concentration of solution applied] ´ [volume of solution applied]. If either are incorrect, results could be unpredictable, unrepeatable, and unacceptable.

*Application Technique*: As with the target tissue, the method of delivering the growth retardant depends on both the chemical used and the plant species to be treated.

*Dips*: With some plants it is possible to dip the plant shoot, or underground portion (bulb, corm) into a growth retardant solution prior to potting. This method is labelled and is very effective for applying B-Nine® to the shoots of rooted chrysanthemums just prior to potting. This is fairly accurate, if each plant remains in the solution for the same amount of time, and if each plant has approximately the same size shoot/ bulb. Unfortunately, this method is not feasible with many of the crops grown, and dipping plants in a common solution could result in the spread of disease organisms.

*Drenches*: Applying a growth retardant in a drench form is fairly easy. Measure out a known amount of chemical, add it to a known volume of water, and apply a known volume of the drench to each pot / plant. Do not apply any plant growth regulator through an irrigation system, unless the label lists chmigation as a legal application technique. Currently, it is legal to apply A-Rest® through drip, overhead boom, sprinkler, and flood subirrigation) systems. Consult the label for specific application recommendations.

*Sprays*: Spray applications can be more difficult to apply evenly. Some chemical labels recommend to "spray to runoff"; that is, spray each plant until spray visibly just begins to drip off of the foliage. Depending on the size of the plant, the sprayer's objectivity, and other factors, varying amounts of chemical will be applied to each plant. It is much safer and more accurate to base spray application on areas, not plants. What does that mean? Apply a known volume of spray to a known area (square footage), regardless of how many plants are in that known area. The general recommendation is to apply ½ gallon per 100 $ft^2$.

If the area is sprayed evenly, it assures that each pot will receive the same amount of spray (the amount is easy to calculate, too), regardless of how many containers are in the area. Unless growers measure out the spray applied to each pot (or calibrate your output rate and time the spray for each pot while maintaining a constant tank pressure), there is no other way to evenly apply the spray over your plants. When addressing height control in greenhouse crops, remember to evaluate all available possibilities and to base decisions on cost, quality, and best management practices.

If chemical control is the most feasible choice, make applications correctly and accurately. Uniformity and consistency are crucial to attain predictable and desirable results. Also remember that growth regulators are regarded as pesticides and that it is a violation of Federal and State Law to use these products in a manner inconsistent with their labels.

# 10

# Carbohydrate Metabolism and Turfgrasses

Based upon their photosynthetic pathways, in cool- and warm-season grasses are referred to as $C_3$ and $C_4$ plants respectively. To establish a basic understanding of the differences between cool and warm-season turfgrasses with respect to their responses to environmental stresses such as drought, temperature extremes, and shade, a familiarity with their differences in photosynthesis and respiration is needed. The balance between the processes of photosynthesis and respiration ultimately controls turfgrass growth, vigour, quality, and resistance to environmental stresses and pests.

### Photosynthesis

Photosynthesis takes place mostly in leaf mesophyll cells, which contain chloroplasts. Chloroplasts are cellular organelles that contain green chlorophyll molecules capable of absorbing the light needed for photosynthesis. Photosynthesis is basically a two-step process whereby green leaves convert solar energy to chemical energy (light reaction) that is used to produce energy-rich carbohydrates (food) using carbon dioxide ($CO_2$) and water ($H_2O$) (dark reaction). The first step requires light, and thus is referred to as the light reaction.

The fixation of $CO_2$ into carbohydrates is light independent, and thus is called the dark reaction (even though the process occurs during the daytime). Radiant energy captured during photosynthesis is stored in the form of chemical bonds in compounds such as adenosine triphosphate (ATP) and as reducing power (source of electrons) in nicotinamide adenine dinucleotide phosphate (NADPH), with oxygen ($O_2$) released as a by-product. Adenosine triphosphate is a high-energy molecule, and the removal of phosphate groups from its structure results in the release of a substantial amount of free energy.

As such, ATP is the major energy currency of the cell and is the driving force for most of its energy-consuming activities. The energy contained in ATP is used to facilitate the transfer of electrons in NADPH to $CO_2$, forming simple carbohydrates that are converted to the sugars, glucose, fructose, and sucrose ($C_6H_{12}O_6$). The photosynthesis reaction is simplified as:

$$\underset{\text{Carbon dioxide}}{6CO_2} + \underset{\text{Water}}{6H_2O} \xrightarrow{\text{light and chlorophyll}} \underset{\text{Carbohydrates}}{C_6H_{12}O_6} + \underset{\text{Oxygen}}{6O_2}$$

### $C_3$ Pathway

A majority of plants assimilate $CO_2$ via the $C_3$ photosynthetic pathway, in which the first stable product formed following $CO_2$ fixation is phosphoglyceric acid (PGA). PGA is a three-carbon compound, hence the name $C_3$ photosynthesis. When the stomata of $C_3$ grasses are open during the day, $CO_2$ enters the leaf and diffuses into the mesophyll cells, where it is fixed in the chloroplasts (Fig. 1). Initially, $CO_2$ reacts with ribulose bisphosphate

(RuBP) to form an intermediate six-carbon compound. This process is called carboxylation.

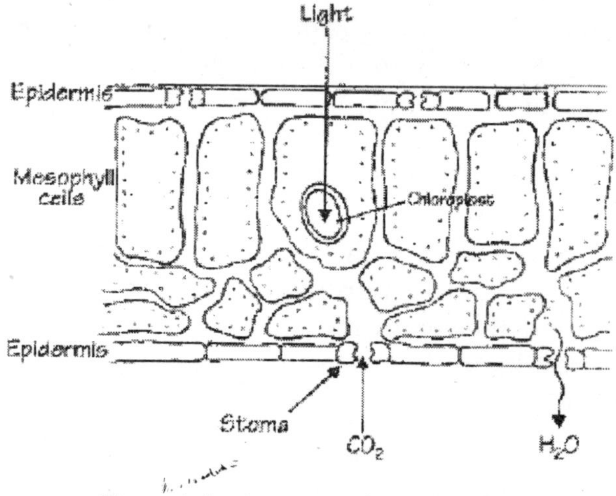


*Figure 1. Leaf structure for $C_3$ turfgrasses*

RuBP is a five-carbon compound that functions as the acceptor molecule for $CO_2$ through a reaction catalysed by the enzyme ribulose 1,5-bisphosphate carboxylase (rubisco). All enzymes, including rubisco, are proteins and act either to facilitate the interaction of two molecules in a reaction in order to produce new molecules or to break one large molecule into smaller ones. Rubisco is the most prevalent and important enzyme in plants. The six-carbon compound formed by the union of $CO_2$ and RuBP is unstable and breaks down into two molecules of phosphoglyceric acid (PGA). What follows is a simplified formula for the $C_3$ photosynthetic pathway, although many additional reactions are involved:

$$CO_2 + \underset{\text{Ribulose 1,5-bisphosphate}}{\text{RuBP}} + H_2O \xrightarrow{\text{Rubisco}} \underset{\text{3-phosphoglyceric acid}}{\text{PGA}} + O_2$$

PGA is then reduced using NADPH and ATP to form a threecarbon sugar used as a building block for larger carbohydrates that are produced in a long series of photosynthetic reactions referred to as the Calvin cycle, Calvin-Benson cycle, or photosynthetic carbon reduction cycle. The final products of $C_3$ photosynthesis are glucose and fructose, which are used to make sucrose and starch. Sucrose, a water-soluble sugar, is transported to sink tissues such as growing leaves, roots, and reproductive structures through the plant via the phloem, and supplies immediately available energy to nonphotosynthesising tissues. Starch consists of long chains of glucose that are not readily available for use in cell metabolism. It is often stored inside cells as a carbohydrate reserve. When plants need to use the energy for cell division or growth, they break down starch into glucose.

#### *Photorespiration*

Rubisco is a dual-function enzyme. It catalyses not only the carboxylation of RuBP (the assimilation of $CO_2$ to RuBP in mesophyll cells) but also the oxidation of RuBP by oxygen molecules (the addition of $O_2$ to RuBP), which initiates a wasteful sequence of reactions collectively called photorespiration. During photosynthesis, oxygen can accumulate to relatively high concentrations in leaves and chloroplasts, stimulating photorespiration. In photorespiration, $O_2$ substitutes for $CO_2$ as a substrate for the rubisco enzyme, leading ultimately to the release of $CO_2$.

In other words, $CO_2$ and $O_2$ are alternative substrates that compete with each other for the same binding sites on rubisco. Whether the carboxylase or oxygenase reaction takes place depends on the availability of $CO_2$ or $O_2$ to the enzyme. At current atmospheric concentrations (300-400 ppm $CO_2$, 21 percent $O_2$), both reactions take place. When the $CO_2$ concentration is high and that of $O_2$ is low, carboxylation is favoured and carbohydrate synthesis proceeds through the Calvin-Benson cycle. However, when the opposite occurs and the concentration of $CO_2$ is low and that of $O_2$ is high, oxidation is favoured and photorespiration results.

The oxidation process breaks down RuBP to PGA and P-glycolate (a two-carbon acid). Two molecules of P-glycolate are further metabolised to $CO_2$ and PGA with the consumption of ATP. Photorespiration is particularly problematic at temperatures above 87°F (30°C). The photorespiration reaction can be simplified to:

$$O_2 + \text{RuBP} \xrightarrow{\text{Rubisco}} \text{PGA} + 2\text{P-glycolate}$$

$$2\text{P-glycolate} + O_2 \rightarrow \text{PGA} + CO_2$$

Carbon fixation through the $C_3$ pathway is most efficient when temperatures are between 60 and 75°F (16 and 24°C). Cool-season grasses are most efficient at fixing $CO_2$ and making carbohydrates when water is plentiful and stomata remain open, allowing $CO_2$ to enter the leaves.

### *$C_4$ Pathway*

Grasses that employ the $C_4$ pathway have a special leaf anatomy, with prominent bundle sheath cells surrounding the leaf vascular bundles (veins) and closely arranged mesophyll cells. $C_4$ grasses also have chloroplasts in both the mesophyll and bundle sheath cells. The initial carboxylation process in $C_4$ grasses is identical to that outlined for $C_3$ grasses. However, $C_4$ plants have evolved an efficient scheme for capturing $CO_2$ and creating high concentrations of $CO_2$ inside bundle sheath cells to react with rubisco in the Calvin-Benson cycle.

As in the $C_3$ pathway, $CO_2$ enters the leaves through the stomata in $C_4$ plants. When absorbed into the mesophyll cells, the $CO_2$ then forms bicarbonate ($HCO_3^-$) in water. Bicarbonate reacts with a three-carbon compound, phosphoenolpyruvate (PEP), to form a four-carbon compound, oxaloacetate, which is then converted into malate in the chloroplasts of mesophyll cells. This process is catalysed by the enzyme PEP carboxylase.

$$HCO_3^- + \underset{\text{Phosphoenolpyruvate}}{PEP} + H_2O \xrightarrow{\text{PEP carboxylase}} \underset{\text{Oxaloacetate}}{OAA} \text{ (4 carbon)} + Pi$$

Malate is next transported to adjacent bundle sheath cells and decarboxylated, causing the release of $CO_2$. A high concentration of $CO_2$ accumulates in the bundle sheath cells for carbon fixation. This released $CO_2$ is refixed by rubisco and assimilated using the enzymes and reactions of the Calvin-Benson cycle, eventually forming sugars. Compared to $C_3$ plants, photorespiration is minimal in $C_4$ plants. Unlike RuBP carboxylase, PEP carboxylase only catalyses the carboxylation of PEP (the addition of $CO_2$ to PEP) using $HCO_3$ in mesophyll cells, and the enzyme does not react with oxygen; hence, there is no competition between $CO_2$ and $O_2$ in this reaction.

In addition, $C_4$ plants avoid photorespiration by locating the carbon fixation steps or Calvin-Benson cycle in the bundle sheath cells, where the rubisco enzyme is saturated with $CO_2$, deterring oxygenase activity and any photorespiration. Furthermore, in the cytoplasm of photosynthetic cells, $CO_2$ is in equilibrium with $HCO_3^-$, and under physiological conditions, this equilibrium results in a $HCO_3$ concentration that is many times greater than that of $CO_2$. Grasses that use the $C_4$ pathway are well adapted to hot and arid climates and exhibit higher photosynthetic efficiency and use water more efficiently than cool-season grasses for the following reasons:

- Optimal temperatures for photosynthesis are lower in $C_3$ than in $C_4$ plants.
- Photorespiration is greater at higher temperatures, reducing carbohydrate production through photosynthesis and negatively affecting energy available to $C_3$ plants.
- $C_4$ plants continue fixing carbon and producing carbohydrates using $CO_2$ concentrated in the bundle sheath cells, even when stomata are closed on hot, dry days. (Once the accumulated $CO_2$ in bundle sheath cells is used, however, carbohydrate production ceases.)
- $C_4$ plants generally have fewer stomata than $C_3$ plants and lose less water while $CO_2$ is being fixed, therefore exhibiting high water use efficiency.

### *Elements of Photosynthesis*

#### *Light*

Photosynthesis occurs only when sufficient light is available. Except for turfgrasses growing in shade, light intensity does not normally limit turf photosynthesis. The photosynthetic rate of $C_3$ plants increases with light intensity to a peak and then plateaus (Fig. 2). The level of light intensity at which photosynthesis reaches a maximum is referred to as the light saturation point (LSP). $C_4$ plants typically have a higher LSP than $C_3$ plants. Some $C_4$ plants do not become light saturated even in full sunlight. Very high light intensities may bleach chlorophyll and retard photosynthesis, but plants that typically grow under such conditions have evolved mechanisms for protection, such as thick leaf cuticles or hairy leaves to reflect light. Plant injury resulting from high light intensity is due not to the light per se but to an excess of light energy over that utilised by photosynthesis.

When light reaching the leaves is not used for photosynthesis, the excess energy triggers production of free radicals that can damage cells (oxidative damage). This often occurs when light intensity is high but photosynthesis is inhibited due to stress from temperature extremes, drought, or excessive soil water. When light intensity is at a low level where photosynthesis and respiration reach equilibrium and the net carbon gain is zero, no plant growth will occur. This light level is the light compensation point (LCP).

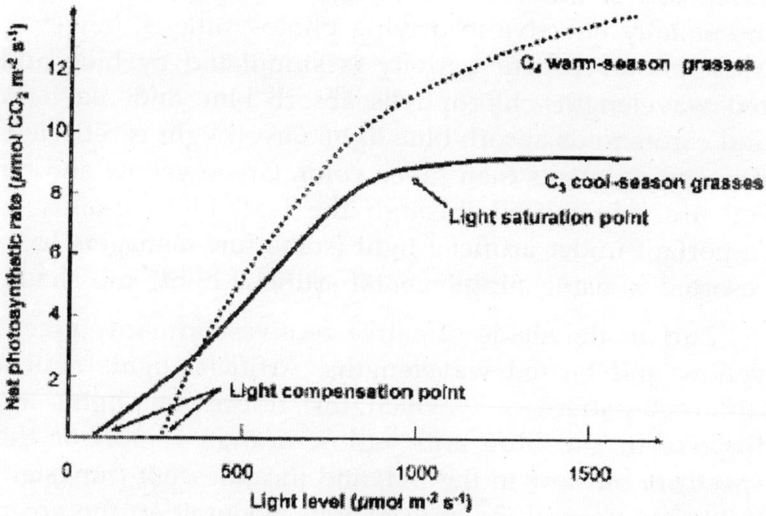


*Figure 2. Photosynthetic responses of cool- and warm-season grasses to increasing light levels.*

Both LSP and LCP vary among turfgrass species and with temperature and $CO_2$ concentration. Under high irradiance, warm-season grasses maintain a higher rate of photosynthesis than cool-season grasses. However, cool-season grasses have a lower LCP and exhibit higher photosynthetic rates under low light levels compared to warmseason grasses. Photosynthetic rates of both warm-season and coolseason grasses exhibit a diurnal pattern on clear, sunny days, increasing from sunrise, reaching a maximum around noon, and then decreasing to the lowest levels by sunset. Photosynthesis is affected by light duration because it occurs only during daylight. Increasing light duration may not increase the rate of carbon fixation, but the total amount of carbon fixed by photosynthesis will increase due to increased light exposure.

Sunlight has all the colours of visible light and is composed of different wavelengths. Not all wavelengths are equally effective in driving photosynthesis, however. Most photosynthetic activity is stimulated by blue and red wavelengths-chlorophylls absorb blue and red light and carotenoids absorb blue light. Green light is reflected, thus giving plants their green color. Green-yellow and far red are transmitted through the leaf. Light quality is important under artificial light (some turf managers have resorted to using supplemental artificial light) and shade.

Turf in the shade of a tree receives primarily green-yellow and far-red wavelengths. Artificial lights emit a different pattern of wavelengths: fluorescent lights are highest in the blue and yellow-orange region of the spectrum but low in the red, and incandescent (tungsten) lights are poor in the blue region, moderate in the green region, and high in the red and far red region of the spectrum, with up to 50 percent of their output in the infrared region.

*Temperature*

Many reactions in photosynthesis are controlled by temperaturesensitive enzymes; therefore photosynthesis is sensitive to temperature change (Fig. 3). Depending on turfgrass species, photosynthesis can occur between 32 and 104°F (0 and 40°C). The optimal temperature range for photosynthesis in cool-season grasses is 68 to 77°F (20 to 25°C ), and for warm-season grasses it is 86 to 95°F (30 to 35°C). It is clear that optimal temperatures for photosynthesis in $C_3$ and $C_4$ grasses are in line with the preferred temperatures for plant growth. As the plant's enzymes function closer to their optimal working temperature in cool-season grasses, photosynthetic rates double with increasing temperature from 50 to 68°F (10 to 20°C).

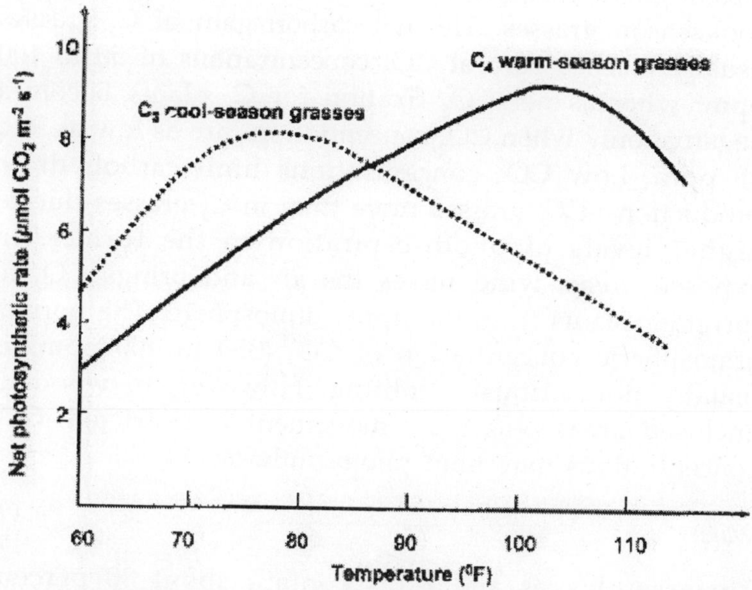


*Figure 3. Photosynthetic responses of cool- and warm-season grasses to increasing temperature.*

When cool-season grasses are exposed to heat stress, photosynthetic rates decline because of greater photorespiration and possible enzyme denaturation. When exposed to full sunlight and high temperatures, $C_4$ grasses are able to produce twice the dry matter that $C_3$ grasses can produce. Carbon dioxide concentration directly affects photosynthetic rates, as $CO_2$ is assimilated into organic compounds in photosynthesis. Photosynthetic rates increase with increasing $CO_2$ concentrations. The optimal short-term $CO_2$ concentration is about 500 ppm, but prolonged exposure to high concentrations can cause damage.

The $CO_2$ concentration at which carbon fixation in photosynthesis is in balance with carbon loss through respiration is called the $CO_2$ compensation point. Warm-season grasses are more efficient at assimilating $CO_2$ than cool-season grasses. The net carbon gain of $C_3$ grasses usually reaches zero at $CO_2$ concentrations of 50 to 100 ppm, whereas net $CO_2$ fixation for $C_4$ plants becomes negative only when $CO_2$ concentrations are as low as 5 to 15 ppm. Low $CO_2$ concentrations limit carbohydrate production in $C_3$ grasses more than in $C_4$ grasses due to higher levels of photorespiration in the former. In exposed areas, wind mixes the air and brings $CO_2$ to turfgrass plants from the upper atmosphere. The normal atmospheric concentration of $CO_2$ (350 to 400 ppm) is usually not seriously limiting. However, in wooded, enclosed areas where air movement is restricted, $CO_2$ concentrations may limit photosynthesis.

### *Water*

Turfgrass leaves and roots contain about 80 percent water, and stems contain about 50 percent water. Photosynthesis is sensitive to changes in plant water status and is inhibited when water deficits or excesses occur. Inhibition of photosynthesis under water stress may be due to stomatal closure and/or metabolic limitations. Stomata control the entry of $CO_2$ into the leaves for photosynthesis. If the plant has too little or too much water, stomata will close and limit the entry of $CO_2$ into leaves, and thus chloroplasts can be deprived of $CO_2$ needed for photosynthesis.

Stomatal closure is often thought to be the first plant response to water stress, and it can occur within seconds. Even a slight water deficit, with no visible signs of wilt, can lead to stomatal closure and a reduction in photosynthesis. Water may also indirectly affect photosynthesis by regulating the synthesis of abscisic acid (ABA), a plant hormone. Abscisic acid production increases in roots in response to a water deficit and is then transported to leaves via the xylem and into the guard cells, inducing stomatal closure. Limitations of photosynthesis resulting from stomatal closure can easily be detected indirectly by a measured increase in leaf temperature due to reduced transpirational cooling.

Plant water status also influences photosynthesis by affecting carboxylating enzymes, electron transport, and chlorophyll content. Ultimately, it is the splitting of water molecules using solar energy during photosynthesis that generates the electrons needed to reduce NADP to NADPH, which in turn reduces $CO_2$ to carbohydrates, driving plant energy production. Lack of water also can directly affect electron transport, which occurs in the light reactions and through which ATP is generated from ADP. Many enzymes are involved in the conversion of solar energy to chemical energy and in carbon fixation. Water deficits and excesses limit the activity of various photosynthesising enzymes, particularly RuBP and PEP carboxylase.

*Availability of nutrient*

Nutrient availability is equally important for photosynthesis in both warm- and cool-season grasses. Nitrogen and magnesium are essential constituents of the chlorophyll molecule, and iron serves as an activator for enzymes involved in chlorophyll synthesis. Sulphur is involved in photosynthetic electron transport and is a constituent of several iron-sulphur proteins that are intermediates in the process, as are manganese and copper. Chlorine is required for photosynthetic oxygen evolution. Lack of nutrients also induces stomatal closure, leading to reduced photosynthesis. Potassium acts as an osmoregulator, and its accumulation promotes the maintenance of cell turgor and stomatal opening, thus affecting photosynthesis.

*Characteristics of leaf*

Leaves contain chlorophyll, which traps the sun's energy, and the photosynthetic rate typically increases with increasing chlorophyll content. Plants with greater leaf area also have higher rates of canopy photosynthesis. Damaged, senescent, or diseased leaves have lower photosynthetic rates than healthy leaves. As leaf angle increases, lower leaves become shaded and total light absorption decreases, resulting in lower photosynthetic rates. Pubescent leaves reflect more light and are less photosynthetically efficient.

### Respiration

Respiration is the process plants use to release the energy stored in the chemical bonds of sugars that are produced during photosynthesis. Warm- and cool-season turfgrass species have the same respiratory pathway. Unlike photosynthesis, respiration takes place in all living parts of a plant, including leaves, stems, crowns, stolons, rhizomes, and roots. Respiration is carried out inside the numerous mitochondria of cells during the day and night. Mitochondria are membrane-enclosed organelles distributed through the cytosol of all plant cells. Their main function is to convert energy from carbohydrates to NADH and ultimately into ATP.

Respiration, like photosynthesis, involves a set of complex chemical reactions that accomplish the breaking and formation of chemical bonds. Respiratory energy is the driving force for metabolism, cell maintenance, growth, and nutrient uptake and transport processes. Aerobic respiration is a biological process whereby carbohydrates (mainly glucose produced in photosynthesis or released when stored starch is hydrolysed) are used to produce energy in the presence of oxygen. The first step in respiration is the splitting of the glucose molecule, through a process called glycolysis, to form two molecules of pyruvic acid.

Energy released in glycolysis is coupled to the synthesis of ATP, the high-energy-storage molecule. For each molecule of glucose that enters glycolysis, two molecules of ADP are converted into ATP. Each molecule of pyruvic acid has the potential to generate eighteen ATP molecules. Thus the two molecules of pyruvic acid yield thirty-six ATPs, plus two from glycolysis, to yield a total of thirty-eight ATP molecules from each molecule of glucose. Chemical energy stored in ATP is essential for plant cell maintenance, growth, and development. The energy production process of respiration can be expressed using the following formula:

$$\underset{\text{(Sucrose)}}{\text{Carbohydrates}} + 12O_2 \xrightarrow{\text{Enzymes}} 12CO_2 + 12H_2O + \underset{\text{(Chemical energy)}}{\text{ATP}}$$

In some cases, such as in flooded or compacted soils, living tissues can respire in the absence of oxygen using anaerobic respiration (fermentation). This does not release as much energy as aerobic respiration, and it produces waste products, such as ethanol, that can be toxic to the plant. This respiration pathway produces only two ATP molecules from each glucose molecule. Therefore, aerobic respiration is nineteen times more efficient than anaerobic respiration. However, if oxygen is not available, anaerobic respiration replaces aerobic respiration to allow glycolysis to continue and keep plants alive, at least for a while.

### *Elements Related to Respiration*

#### *Carbohydrate*

Respiration rate depends on the availability of carbohydrates for enzyme function and energy production. Plants with low carbohydrate reserves cannot support high respiration rates for long. When the carbohydrate supply is exhausted, proteins can be used in respiration, but this will limit plant growth or result in no net growth. Leaves receiving full sunlight have higher respiration rates than shaded leaves, probably because there are more sugars available as a result of higher rates of photosynthesis. In fact, leaf respiration rates are much higher just after sunset, when sugar levels are high, than just before sunrise, when sugars have been consumed overnight.

### *Oxygen*

Oxygen is required for plants to carry out aerobic respiration and for efficient energy production. Ambient air contains a sufficient oxygen supply for aerobic respiration under normal conditions. The rate of oxygen penetration or diffusion into leaves, stems, and roots is usually adequate to maintain normal intracellular levels. Low ambient $O_2$ supplies can occur in flooded, saturated, frozen, or compacted soils. Under such conditions, aerobic respiration is inhibited, resulting in low energy production, and plants may employ anaerobic respiration. Low oxygen levels are less detrimental for respiration and growth under low temperatures (as in winter) than they are under high temperatures (as in summer) due to differences in the demand for $O_2$.

### *Temperature*

There is an optimal temperature for respiration in all turfgrass species, with a higher optimum for warm-season than for cool-season grasses. Respiration ceases at temperatures near freezing and increases exponentially with temperature. Extremely high temperatures eventually result in reduced respiration as substrates become exhausted, enzymes become denatured, and tissues deteriorate. For most plant species, the respiration rate doubles for every 18°F (10°C) rise in temperature between 41 and 77°F (5 and 25°C).

### *Water*

Respiration requires many enzymes that do not function if the plant is under stress from either too little water or too much. Lack of water restricts carbohydrate availability, in part because water acts as a solvent for carbohydrates during transport and metabolism.

Excessive soil water inhibits root respiration by reducing oxygen diffusion rate and availability. Oxygen diffuses ten thousand times more slowly in water than in air. Low root respiration rates are common in waterlogged or frozen soils.

*Nutrient*

Nitrogen stimulates respiration due to an increased demand for growth, ion transport, and nitrogen assimilation. Nitrate application can result in higher root respiration rates relative to ammonium, possibly due to the greater demands for metabolic energy required for nitrate uptake and reduction to ammonia. Phosphorus is a component of enzymes and proteins involved in respiration and a major constituent of the energy transfer molecule ATP. As such, respiration rates are lower when a phosphorus deficiency exists.

*Physical conditions of plants*

There is generally a positive correlation between growth rate and respiration rate. Younger tissues or plants have higher respiration rates than older ones. Root tips and shoot meristems have high respiration rates. Senescent leaves still respire for limited maintenance but have a low rate of growth respiration. Much of the respiration in mature plants occurs in young leaves and roots and in growing flowers.

***Balance between Photosynthesis and Respiration***

Respiration involves carbohydrate consumption and energy release, whereas photosynthesis is a carbohydrate production and energy absorption process. Between 25 and 75 percent of all the carbohydrates produced per day are respired by the plant, depending on turfgrass species and growing conditions. Plant carbohydrates that are not used for respiration will be translocated to nonphotosynthesising storage organs (such as the crown, stems, stolons, rhizomes, roots, and seeds) or stored temporarily in the photosynthesising tissues (leaves).

Sucrose is the most common carbohydrate transported in plants. Starch and fructans, on the other hand, serve as storage carbohydrates. Fructans are the predominant storage carbohydrate in cool-season grasses, whereas starch is the most prevalent in warm-season grasses. Stored starch or fructans are broken down into more usable forms (sucrose or glucose) when carbohydrate supplies are low. Glucose, fructose, sucrose, and fructans are water-soluble carbohydrates (WSCs), and they accumulate in leaves and stems.

Water-soluble carbohydrates are readily available for use in metabolism and growth. Total nonstructural carbohydrates (TNCs) are composed of WSCs and starch and are often used as a measure of total carbohydrate reserves. When plants are actively growing or producing flowers and seeds, the demands for energy are high, and stored carbohydrates are used as an energy source, resulting in a decline in the carbohydrate concentration. If growth is slow and the demand for energy is low while photosynthesis continues at a rapid rate, carbohydrates will accumulate.

Maintenance of a positive carbohydrate balance is important for plant health. This results when the net accumulation of carbohydrates from photosynthesis exceeds consumption in respiration. A positive carbohydrate balance allows plant growth to continue after the plant's maintenance needs have been addressed. The plant demands more carbohydrates for growth than for maintenance or storage. When carbohydrate production is insufficient to meet plant needs, the stored carbohydrates are used for growth until they are exhausted. At that point, growth will decline, and the plant may die from carbohydrate starvation, although metabolic dysfunctions will probably prove lethal before carbohydrates are exhausted.

Any environmental condition or cultural practice that reduces carbohydrate production in photosynthesis or increases carbohydrate consumption in respiration to a level where carbohydrate depletion occurs can have a negative effect on plant health. For example, nitrogen fertiliser application has long been known to reduce carbohydrate levels in leaf and crown tissues. The decline in reserve carbohydrates in response to nitrogen may result from the increased shoot growth and respiration that are stimulated by its application. Respiration has a higher optimal temperature than photosynthesis.

High temperatures increase respiration rates and can lead to carbohydrate depletion, particularly when photosynthesis starts to decline. WSC content increases during the day, when photosynthesis takes place, and decreases during the nighttime due to respiration. Long exposure of turfgrass to shaded conditions and persistent low mowing heights reduces canopy photosynthetic capacity while maintenance respiration continues, and this can result in carbohydrate depletion.

An adequate supply of storage carbohydrates is extremely important for winter survival and spring regrowth of perennial grasses, as leaves senesce and photosynthetic capacity is much reduced in cool-season grasses and nonexistent in warm-season grasses during the winter. In warm-season grasses, carbohydrate reserves are necessary for plant respiration during winter dormancy, when photosynthesis is not possible but crowns and roots remain alive. Therefore, proper turfgrass management in the fall must maximise carbohydrate production and avoid depletion of food reserves during winter.

# 11

# Hormones, Signals and Target Cells in Plant Development

In multicellular organisms, communication among neighbouring cells and with distant parts of the growing body is crucial to achieve coordinated development and healthy growth. Vascular systems play an essential role both in delivering nutrients, oxygen, etc. and in long-distance signalling for developmental processes.

According to Darwin's studies of tropisms in plants, informational signals passed from one part of the plant to another, the proof that it was a chemical substance that passed awaited the famous Avena coleoptile experiments of Frits Went. These showed that a molecule (later identified as indole-3-acetic acid [IAA]) was the active agent that was water soluble and would pass across an agar barrier placed between one tissue and another—in his earliest experiments this was between the coleoptile tip (producing IAA) and the IAA-regulated elongating region of the coleoptile below.

A tremendous amount of work, both in studying the physiology of this response to IAA, and in identifying the many analogues to IAA, sought the molecular structures required to provide an active molecule. It was from this highly intensive period of plant physiology study that the agricultural revolution of herbicides, defoliants and growth regulators of the 1940s and 1950s was originally generated. But it was the insect physiologists with their identities of hormone-producing glands and hormone-responding tissues remote from the glands who developed the concepts of target tissues, signalling molecules and receptor sites. Perhaps the most spectacular to record, as an example of the approaches followed later by plant scientists, is the work in the 1930s and 1940s concerning the processes of moulting of larval epidermal skins and of metamorphoses to the adult state.

The prothoracic gland produces the steroid hormone ecdysone (the moulting or juvenile hormone) that is transported in the haemolymph to the insect epidermis. Along the way, ecdysone binds to specific proteins and then at the site of moult, induces enhanced transcription and modification of coordinated gene activity. Importantly, ecdysone-induced puffing at specific sites in polytene chromosomes could be demonstrated in salivary glands and related to each developmental stage of differentiation. The insect field was set for determining the cascade events of signals, receptors and the hormonal regulation of gene expression in specific target cells.

The plant field followed fast; but in the absence of specific glands or polytene chromosomes and the absence of distinguishably specific target cells the now recognised parallels between plant and animal signalling were more difficult to explore and took longer to resolve. The concept of target cells in plants arose originally from the knowledge that cells of coleoptiles and etiolated shoots would enlarge and extend in response to auxin and that ethylene would arrest such elongation growth. Those of us who worked in the field of abscission became equally aware that the cells that made up an abscission zone enlarged prior to their separation and did so with alacrity if exposed to ethylene, but if given auxin these same cells would neither grow nor separate.

The insect developmental biologists were aware that only certain cells would perceive and respond to particular hormonal signals in a particular way. It became inevitable to learn that although most plant cells might look the same, they were all as individual and distinct from one another as those making up the highly responsive organs of mammals and insects. As the cuticular cells of a larva were targets for the moulting hormone ecdysone, and the mammalian liver for insulin, so the cortical cells of the plant shoot were targets for auxin and ethylene. The concept for target cells in the developmental biology of plants was born.

A first substantive evidence for different target types in cortical cells arose from observations of the differences in their growth responses to auxin and ethylene. Whereas the immature cells of young dicotyledonous shoots such as those of the pea *Pisum sativum* will elongate in the presence of the auxin passing downwards from the meristem, the addition of ethylene to those shoots will cause them to arrest elongation growth and instead to expand laterally, with the cell volume remaining essentially unchanged and determined by the availability of auxin from the meristem. Cells of abscission zones, however, such as those in the leaves of the bean *Phaseolus vulgaris*, behave in a quite different way: their expansion is enhanced by ethylene but not by auxin.

The existence of a third type of cortical cell is found in the stems and petioles of many species of flooded or aquatic habitats. Ranunculas sceleratus or the water fern Regnellidium diphyllum, for example, possesses cortical cells that will expand and extend with either auxin or ethylene (Figure 1). In 1976, these three distinguishable cell types were designated as Type 1, Type 2 and Type 3 with respect to their responses to auxin and ethylene. Once we understood that cells that looked similar to the eye had quite specific responses to hormonal signals, it became evident that strict regulatory controls operated to maintain these coordinated patterns in the cell society.

Clearly, not every cell differentiated along the same developmental pathway or had the competence to respond in a similar way to the same hormonal signals. More examples of target cells other than those with a highly specific perception and response to auxin and ethylene signals have now engaged the physiologists and molecular biologists, none less so than the terminally differentiated cells of the aleurone tissue in graminaceous seeds. In aleurone cells, the competence to respond to a gibberellin/abscisic acid control determined their final response and cell fate.

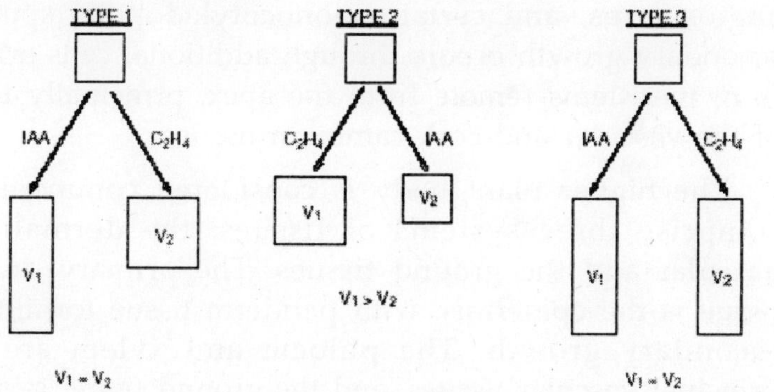


*Figure 1. Overview of Type 1, Type 2 and Type 3 target cells in terms of their responses in planta to auxin and ethylene.*

The terms target cell, signal, cellular competence and tissue sensitivity are now used widely when describing the development of eukaryotes, but differences have emerged in terms of the definitive definitions of their meaning.

### Target Cell

The term target cell was first conceived in animal biology and is generally taken to mean a cell that has a pre-determined competence to respond in a defined way to a specific hormonal signal. Only the specific hormone (or a chemically related analogue) can evoke that particular response in the cell. In plant biology, the definition is essentially the same.

### Cells and Tissues

The body of higher plants arises from the mitotic activity of apical meristematic regions, and thence through cell elongation and differentiation the primary plant body is formed. In the gymnosperms and dicotyledonous angiosperms, and certain monocotyledonous species, secondary growth occurs through additional cells arising from meristems remote from the apex, principally those of the vascular and cork cambium tissues.

The higher plant body is considered commonly to comprise three systems of tissues: the dermal, the vascular and the ground tissues. The primary dermal tissue is the epidermis, with periderm tissue forming in secondary growth. The phloem and xylem are the principal vascular tissues, and the ground tissue systems comprise all the remaining tissues, with parenchyma the primary cell type. Within these tissues, a further layer of complexity emerges with the organisation of cell types. In some tissues, only a single cell type is found and these are referred to as simple tissues. In others, for example the stele, many cell types comprise the tissue, including those that are dead, and these are commonly referred to as complex tissues.

### Signals

Within the plant body, cells must, perforce, communicate by chemical or physical means. These intertissue messages are the chemical signals that evoke specific biochemical and molecular events in each responsive target cell. From the first of these signalling molecules to be discovered, namely auxin in 1928, a wide spectrum of such molecules has now been identified and while the chemistry of these compounds is quite diverse they all exhibit certain shared characteristics. Although there are major sites of synthesis within the plant, signals all move readily between cells (either long or short distances) and evoke responses both at the site of synthesis or at sites that are remote.

Classically, for a cell to recognise a signal, it must have a receptor for that signal. All cells must, therefore, possess an array of receptor systems, but so far comparatively few of these have been precisely characterised. However, the identification and characterisation of these receptors is currently a very active research area in plant biology and as such represents areas of high relevance to the target cell concept.

The five major classes of plant hormones (auxin, ethylene, abscisic acid, gibberellin, cytokinins) are well established, but many more endogenously produced molecules with equally diverse structures are now known to induce a range of specific responses in specific tissues. These include:

– steroid-like molecules, e.g., the brassinolides,

- low molecular weight compounds such as jasmonic acid and salicylic acid,
- oligosaccharins, including free N-glycans, and
- peptides, such as systemin.

### *Cellular Competence*

A cell must possess the ability to perceive, transduce and respond to a signal, otherwise it is not a target for that particular input. In plants, unlike animals, all the living cells are exposed to the same hormones, though the signal transduction pathway is not necessarily the same for each target cell. Whereas a positive response to the signal is evidence of a cell's specific target state, lack of response cannot be taken as evidence of a non-target condition. For in every signal input there must be a threshold level that a cell can interpret; this must exceed the presence of non-specific inhibitors and homeostatic controls exerted by specific repressor controls of the target cell response. Furthermore, a cell may perceive, transduce and respond to a particular signal in a different way from its neighbouring cell.

### *Cell and Tissue Sensitivity*

A highly sensitive target tissue requires a low concentration of signal input to evoke a predetermined response—i.e., display its tissue competence. It should be noted that any one cell may be a target for more than one signal; therefore the sensitivity of the tissue for each signal may vary, and must be defined in terms of a specific signal. The identification of definitive receptors for some of the plant hormones—for example, the ETR-like family of proteins as ethylene receptors—has afforded the possibility of quantifying sensitivity in terms of receptor abundance. Nevertheless, researchers who have attempted such exercises are still bound to equate such values with the extent of the physiological response.

### Long-range Signalling

Communication between distant organs is especially relevant in plants due to the fact that they cannot escape from adverse conditions. In order to survive and perpetuate, they have to adapt their growth and development to the changing environment. Non-cell-autonomous signalling is involved in many plant developmental processes, including shoot morphogenesis, shoot branching, root patterning, flowering and tuber formation.

Evidence of long-distance signalling in plant development has come from several observations. The adult parts of a plant derive from the activity of two groups of stem cells, called meristems, situated at the tip of the root and at the apex of the shoot. Development can be strongly influenced by environmental stimuli, which are often perceived by a plant part distant from the place, usually a meristem, undergoing the developmental response. Shoot development is influenced by stresses affecting the roots, suggesting that roots send signals to the shoot.

Flowering and tuberisation are regulated in many species by photoperiod, the relative duration of light and darkness during a day. Because the main site of photoperiod perception is the leaf, photoperiodic induction of flowering and tuberisation requires transmission of signals from the leaves to either the shoot apical meristem, where flowering takes place, or the stolons, underground stems where tubers develop (Fig. 2). Long-distance messengers regulate flowering in photoperiod-insensitive plants as well, indicating that internal signals can also trigger the production of mobile molecules. Grafting studies have provided proof for the existence of long-range signals in the regulation of flowering time, tuber induction, shoot branching and effects of roots on the shoot.

To understand long-distance signalling in flowering and tuberisation it is convenient to distinguish three successive steps:

1) Induction in the leaves, which determines the timing of the process;
2) Commitment (also called evocation or determination) to flowering or tuberisation at the shoot apical meristem or at the stolon tip, respectively; and
3) Tuber initiation at the stolon tip or floral morphogenesis at the shoot apical meristem.

Thus, export of inducing signals from the leaves follows induction and precedes commitment. An alternative mode of commitment exists: leaves can become committed to continually produce the long-range signal. Defoliation experiments indicate that a particular line of Impatiens balsamina is irreversibly committed to flowering because leaves exposed to inductive photoperiods produce a transmissible signal incessantly. In other plants, like Chenopodium rubrum, Sinapis alba and Arabidopsis thaliana, leaf removal at different times indicates that commitment to flowering occurs at the shoot apex and, therefore, after the leaves have started to export the signal.

Loss of inductive conditions, however, can lead to partial or total reversion of the flowering process in soybean (Glycine max), some lines of I. balsamina and in other species, as well as of tuberisation in potato (Solanum tuberosum). This suggests that in these plants real commitment to flowering or tuberisation does not take place, or occurs after floral or tuber development has already begun. In some species, therefore, these developmental processes are only completed if continuous inductive conditions are maintained. All these situations illustrate the wide variety of responses that can be found in nature.

Although it was initially proposed that flowering is triggered in inductive photoperiods by a single substance, which was named florigen, experimental results support that several transmissible components participate in the regulation of flowering onset. Numerous grafting experiments have shown that leaves exposed to favourable or unfavourable photoperiods produce mobile substances capable of promoting or inhibiting flowering, respectively. Evidence for positive and negative transmissible regulators of tuberisation has also been obtained. Furthermore, long-range signals are interchangeable between plants with different photoperiodic responses.

A good example is the acceleration of flowering in a day neutral tobacco cultivar under short day (SD) conditions when is grafted onto a tobacco cultivar that responds to SD and under long day (LD) conditions when it is grafted onto Nicotiana sylvestris, which responds to LD. Induction of tuberisation by flowering signals has also been described. Tobacco scions kept under conditions inductive for flowering can promote tuberisation when grafted onto potato stocks, whereas non-induced tobacco scions do not cause tuberisation.

An interpretation of these results is that long-distance signals for flowering and tuberisation are identical or very similar for many plants, or even that the same signals are used by all higher plants, as has been pointed out by several authors. There is, however, a number of cases in which interspecific transmission of the signal could not be demonstrated, suggesting that the floral stimulus might not be universal. On the other hand, lack of transmissibility between species could be explained by different amounts of signals formed in the leaves of different species, by different balances between positive and negative regulators, by different sensitivity of the shoot apex to the stimulus, or by rapid reversion to vegetative growth in some instances. As long as the chemical nature of the signals is not determined, it is impossible to establish whether these signals are universal or not.

### Graft-transmissible Signals

Applying inductive photoperiods only to leaves can promote flowering. This has led to the general assumption that mature leaves are the most effective site for photoperiod perception and generation of long-distance information substances. In several species, however, fully expanded leaves are not required for sensing photoperiod. Defoliation experiments indicate that cotyledons of C. rubrum and I. balsamina seedlings produce a floral stimulus that is exported to the shoot apex under inductive SD. Given that Arabidopsis is committed to flower at about 7 inductive LD, before any leaf has reached maturity, photoperiod must be perceived either by cotyledons or by very young leaves.

In maize, the flowering signal is also produced by immature leaves. According to Chapman, photoperiod is perceived by young rapidly expanding leaves in potato, since plants whose terminal leaf cluster is exposed to a particular photoperiod tuberise like whole plants exposed to the same conditions. Long-distance signals that regulate the initiation of flowering and tuberisation move through the phloem, which has two basic cell types, enucleate sieve elements and companion cells.

Any transported molecule has to enter the sieve elements (phloem loading), move with the phloem stream and then leave it (phloem unloading). It is clear that longdistance signalling can have regulatory mechanisms at many levels: synthesis of signals, transport to the vascular bundles, phloem loading, translocation in the phloem sap, phloem unloading, entry into the target tissues and sensitivity of the target cells to the signals.

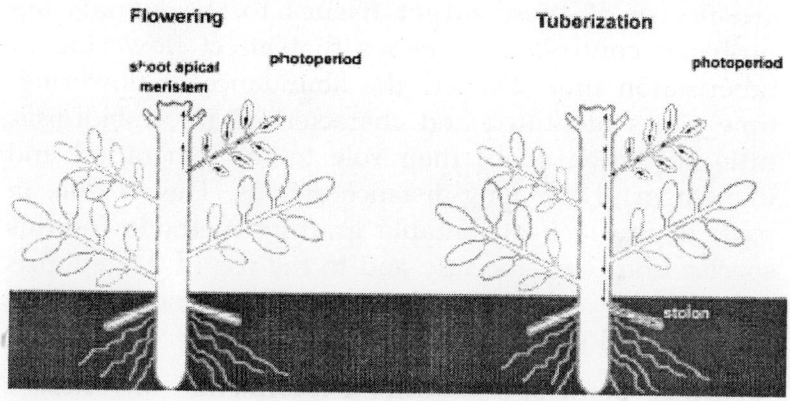


*Figure 2. Movement of long-range signals for flowering and tuberisation.*

In potato, the factors that induce tuberisation are found throughout the entire plant and the signal can move both basipetally (from apex to base) and acropetally (from base to apex) (Fig. 2). However, tuber formation always occurs at the physically lowest nodes of the plant, even if a stem cutting is planted in an inverted position, suggesting a strong effect of gravity either in the accumulation of the stimulus, or in the competence to respond to it. For flowering, the stimulus has to be transported at least acropetally, since flowering occurs at the shoot apical meristem (Fig. 2). In several species, including tobacco, orchids and Arabidopsis, there is a gradient in the potential to flower, decreasing from the apical nodes. It is not known whether this gradient is caused by differential movement or accumulation of the floral stimulus, or different sensitivities of the axillary meristems to the mobile signals.

### Mobile Signal

The production of positive and negative components of the mobile signal, the balance between them, their transport to the shoot apex or the stolon tip and the sensitivity of these target tissues to the signals are probably controlled by genes that affect flowering or tuberisation time. Despite the abundance of flowering-time genes identified and characterised in Arabidopsis, little is known about their role in the generation and movement of the long-distance signals. This is due, in part, to the lack of suitable grafting methods for this species until very recently and to the use of Arabidopsis mainly for genetic and molecular rather than physiological studies.

Other plant species offer experimental advantages that have allowed some progress in understanding the role of several genes in leaf-to-apex communication. In maize, the INDETERMINATE1 (ID1) gene is required to promote the transition to reproductive growth, as id1 mutants flower later than wildtype plants. The analysis of an id1 mutant caused by a transposon insertion in this gene revealed that plants containing sectors in which the transposon had excised flowered earlier than fully mutant id1 plants, even though the revertant sectors did not affect the shoot apical meristem. This indicates that ID1 acts in a non-cell-autonomous manner.

Since this gene is expressed only in immature leaves, it must regulate the production or transport of a signal exported to the shoot apex to induce flowering. Several genes are involved in the generation of a graft-transmissible floral inhibitor in pea and at least one in the production of a floral stimulus. In this plant, the positive signal seems to be produced constitutively. Flowering is thought to occur when the ratio of stimulus to inhibitor exceeds a certain threshold. This is achieved under LD because the photoreceptor phytochrome A reduces the production or transport of the inhibitor.

The gigas mutant of pea, affected in the production of the constitutive floral stimulus, has a phenotype similar to that of Arabidopsis mutants classified in the autonomous (or constitutive) floral-promotion pathway, with late flowering and enhanced photoperiod and vernalisation responses. In contrast, pea mutants affected in inhibitor production—the sterile nodes (sn), die neutralis (dne) and photoperiod (ppd) mutants—flower early in a photoperiod-insensitive manner.

Despite being similar, the phenotype of the pea phytochrome B (phyB) mutant is not graft-transmissible, suggesting that the level of a long-distance floral inhibitor is not altered. The isolation of the genes affected by the gigas, sn, dne and ppd pea mutations might be extremely valuable to identify mobile floral signals. Work on potato has also provided information about genes involved in the production or movement of long-range signals. In this plant, tuberisation is promoted in SD and PHYB represses tuber initiation under non-inductive LD conditions.

Potato plants that carry an antisense PHYB transgene tuberise in both photoperiods. When antisense PHYB scions were grafted onto wild-type stocks, two different results were obtained: if all of the leaves in the stocks were removed, these formed tubers in LD, but when wild-type leaves were left, tuberisation did not occur. Wildtype scions grafted onto antisense PHYB stocks inhibited tuber induction in LD. These results are compatible with two hypotheses: either PHYB is required for the production of a graft-transmissible inhibitor of tuberisation or PHYB is required for phloem loading of this inhibitor in the leaves.

The tuberisation phenotype of antisense PHYB potato plants is comparable to the flowering phenotype of pea phyB mutants, with photoperiod insensitivity and earliness. However, as mentioned above, a transmissible inhibitor is not affected in the pea phyB mutant. CONSTANS (CO) is a central gene in the photoperiodic regulation of flowering in Arabidopsis. CO encodes a nuclear zinc-finger protein required to promote flowering in inductive LD. Expression of this Arabidopsis gene in potato delays tuberisation under SD conditions, which promote tuber formation in wild-type plants.

Grafting experiments indicate that CO probably acts in the leaves, rather than in the stolons, to delay tuberisation of CO-overexpressing potato plants, suggesting that CO acts upstream of the graft-transmissible signal. Two research groups have recently shown that this is the case in Arabidopsis, placing the floral stimulus within the genetic network that controls flowering time. Fusions of the CO promoter to the β-glucuronidase (GUS) reporter gene revealed GUS expression in the shoot apex, young leaves and the vascular tissue of the hypocotyl, cotyledons, leaves and roots.

Expression of CO from several phloem-specific promoters was sufficient to complement the late-flowering phenotype of a co-mutant in LD and to accelerate flowering in a wild-type background in SD. However, expression from meristem-specific promoters did not complement a co mutation, indicating that CO acts in the phloem to trigger flowering. Furthermore, when co mutant scions were grafted to wild-type stocks or to stocks expressing CO from a phloem promoter, the scions flowered earlier than co mutant controls.

Because the CO protein and mRNA do not seem to move from the phloem, these results strongly support that CO regulates the production or entry into the phloem of a long-distance flowering signal. Within the complex genetic network that controls flowering time in Arabidopsis, the functions of numerous genes converge in the regulation of a few genes called floral pathway integrators, which include FLOWERING LOCUS T (FT) and SUPPRESSOR OF OVEREXPRESSION OF CONSTANS 1 (SOC1), also called AGAMOUS-LIKE 20. These two genes are direct targets of CO and ultimately affect the expression or activity of a set of genes termed floral meristem identity genes, which convert a vegetative meristem into a reproductive meristem that produces flowers.

LEAFY (LFY) and APETALA1 (AP1), among others, are floral meristem identity genes. LFY is at least in part downstream of SOC1, and AP1 has been proposed to act downstream of FT, which does not affect LFY expression. Another gene that acts at a similar level to FT and that affects AP1 but not LFY function is FWA. SOC1, FT and FWA might also affect other floral meristem identity genes. Since CO probably regulates the synthesis of the long-distance signal and the floral meristem identity genes act in the shoot apical meristem and therefore after the signal has reached this meristem, FT and SOC1 are good candidates to be involved in signal transport.

Alternatively, the FT and SOC1 gene products might be part of this transmissible signal. FT expression is detected in the vasculature, it can be specifically induced by CO in the phloem and it can promote flowering when directed by different tissue-specific promoters, including phloem and meristem promoters. LFY probably has additional roles besides being a floral meristem identity gene. Many authors include LFY among the floral pathway integrators.

LFY expression is upregulated upon floral induction and LFY in turn upregulates AP1 expression. The increase in mRNA levels of these genes requires the activity of TERMINAL FLOWER 1 (TFL1). TFL1 shows sequence similarity to FT and animal phosphatidylethanolamine-binding proteins. FT and TFL1 regulate flowering in an antagonistic manner, since ft mutants flower later than the wild type and tfl1 mutants are early flowering.

Tfl1 mutants are committed to flower earlier than the wild type, indicating that either the mobile signal is produced earlier, or is transported more rapidly, or the meristem is competent to respond to it earlier. All these alternative hypotheses associate TFL1 very closely with the generation, transport or response to the floral stimulus. Using the newly developed micrografting method in Arabidopsis, it should be relatively easy to test whether the phenotype of the tfl1 mutant is graft-transmissible. It had been suggested previously that TFL1 might influence the response of the apex to mobile signals.

Interestingly, a TFL1 homologue from pea, LATE FLOWERING, controls the sensitivity of the apex to floral

*Table 1. Putative Compounds of Long-Distance Flowering Signals*

| *Compound* | *Supporting evidence* | *Evidence against* |
|---|---|---|
| Gibberellins | Detected in phloem and xylem sap | Opposite roles in different species |
| | Increase at shoot apical meristem after increase in leaves shortly after transfer to inductive conditions | Not sufficient to induce flowering in several plants |
| | Transported from leaves to shoot apical meristem | |
| Cytokinins | Levels increase at shoot apical meristem a few hours after induction | Not sufficient to induce flowering when applied exogenously |
| Sucrose | Levels increase rapidly in leaf exudates under inductive conditions | Difficult to distinguish between signalling and metabolic effects |
| | Correlation between reduced sucrose levels in leaf exudates and late floral induction in a starchless mutant | |
| | Complements the late-flowering phenotype of co mutants, affected in the levels or phloem loading of the flowering signal and of a starch-deficient mutant | |
| Proteins and peptides | Differences in proteins/peptides present in phloem exudates between induced and non-induced plants | Effect of these proteins/peptides on flowering not tested so far |
| | | Phloem exudates collected long after transfer to inductive conditions |

| | | |
|---|---|---|
| mRNAs | Role in long-distance signalling for leaf development<br>Present in phloem sap | Effect of mobile RNAs on flowering not tested so far |
| miRNAs | Present in phloem<br>Easily exportable<br>Sequence specific<br>Affect flowering time | Effect of miRNAs present in phloem sap on flowering not tested so far<br>Presence of miRNAs affecting flowering in the phloem not tested so far |
| Salicylic acid | Promotes flowering in Arabidopsis<br>Involved in long-distance signalling in systemic acquired resistance | Role in long-distance signalling for flowering not tested so far<br>In systemic acquired resistance, it probably activates the synthesis of mobile molecules |

signals, is a repressor of flowering and its effects are not graft-transmissible, giving support to the hypothesis of a similar role for TFL1. The presence of the TFL1 mRNA below the shoot apical meristem is consistent with this hypothesis. Very recent work shows that grafting late-flowering gigantea (gi) mutants to wild-type Arabidopsis stocks accelerates significantly flowering of gi, a result not surprising since GI regulates CO expression.

The flowering time of the ft mutant, however, was only slightly reduced when grafted onto wild-type stocks. One interpretation of this result is that the wild-type stock can provide FT function to the mutant scion, either because FT regulates the synthesis of a mobile signal or because the product of this gene is transported to the shoot apex. Nevertheless, since the acceleration of flowering in ft scions is small, another interpretation is that FT is required for effective movement of the signal.

Further investigation is necessary to establish the role of FT in long-range signalling. The use of the Arabidopsis micrografting assay should allow examining whether other flowering-time genes play any role in the production or translocation of mobile signals.

The existence of long-distance signals for flowering and tuberisation has been known for several decades. Different mol-ecules have been proposed as components of the floral stimulus and inhibitor, but there is no conclusive proof so far that any of them is the flowering signal transmitted from the leaves to the shoot apical meristem. The main compounds postulated as grafttransmissible floral regulators are explained below and summarised in Table 1.

### Plant Hormones

Several plant hormones are thought to regulate flowering

by moving from the leaves to the shoot apex, based mainly on two approaches: the effect of mutations in genes affecting hormone synthesis or hormone signal transduction on flowering and the effect of exogenous applications of hormones or hormone inhibitors on flowering.

The effect of GAs on flowering time has been extensively reviewed recently. Therefore, only a summary of the most significant data is presented here. It must be emphasised that different plant species respond in different ways to the external application of GAs, making difficult to draw general conclusions. GAs can be detected in phloem and xylem saps, indicating that they can be transported. In Arabidopsis, GAs promote flowering, especially under SD conditions.

GAs activate the expression of LFY and SOC1 in Arabidopsis and of SaMADS A (the SOC1 homologue) in S. alba. In several LD and SD species, GAs do not induce flowering unless accompanied by other treatments and in others, GAs inhibit flowering. The species in which the role of GAs as floral stimulus is best documented is Lolium temulentum. In this plant, a single LD is sufficient to promote flowering and this treatment induces a rapid increase of GA20 in the leaves. GA20 is a precursor of bioactive GAs, like GA1, GA5 and GA6.

A few hours later, there is a peak of GA5 and GA6 at the shoot apex. Application of GA5 and GA6 to leaves promotes flowering in SD and LD, showing an additive effect with LD. When labelled GA5 was applied, transport to the shoot apex could be observed. These results strongly support that GA5, and perhaps GA6, might be part of the long-distance signal. However, whether the changes in GA levels at the shoot apex are the result of de novo synthesis or transport from the leaves is not known, because an increase of GA5 and GA6 in the leaves has not been documented so far. Although these results support that GAs affecting flowering can move from the leaves to the shoot apex, it is not clear to what extent these results can be extrapolated to other plant species.

In fact, GAs do not seem to be a major flowering signal in S. alba. Endogenous GA levels also increase in Arabidopsis plants transferred from SD to LD. However, the GAs analysed affect stem elongation and it has not been shown whether they are involved in flowering. Furthermore, movement of these hormones might seem unnecessary because GA biosynthetic enzymes are expressed in many plant parts. In potato, exogenous application of GAs has an inhibitory effect on tuberisation, whereas treatment with GA biosynthesis inhibitors promotes tuber initiation.

Plants with reduced GA levels can tuberise earlier than wild-type plants in SD and LD. Tuberisation in SD is thought to be mediated by a decrease in GA levels, since transfer of plants from LD to SD reduces GA abundance. Although a role for GAs in regulating tuber induction is supported by all these data, the contribution of these hormones to long-range signalling in this process has not been investigated. Several species can be induced to flower by exogenous cytokinins, but in most cases only when the treatment is combined with other factors slightly inductive for flowering.

Levels of these hormones increase in the phloem of white mustard (Sinapis alba) and Arabidopsis plants induced to flower by one LD. A few hours later, the level of specific cytokinins is higher in the shoot apical meristem of induced plants than in non-induced plants. This extra cytokinin correlates in time with an increased flux of this hormone directed to the shoot apex in S. alba and with movement of the floral stimulus to the apex in Arabidopsis.

Although cytokinin application to the apical bud of non-induced S. alba plants causes events related with the induction of flowering, such as stimulation of cell division and expression of certain genes, it is unable to promote flowering. Therefore, cytokinins are involved in the flowering process and there are changes in cytokinin fluxes at the time of floral induction in several plants, but the role of these hormones in long-distance signalling for flowering is unclear. Although an effect of cytokinins and abscisic acid (ABA), another plant hormone, on tuber induction has been suggested, the results obtained until now are not consistent.

### *Sucrose*

Together with GAs, sucrose is the most extensively studied compound that might participate in long-range signalling for flowering. In S. alba and Arabidopsis plants exposed to inductive photoperiods, sucrose levels increase rapidly and transiently in phloem leaf exudates. According to defoliation experiments, this increase coincides with the start of mobile signal transport and occurs before the activation of cell division in the meristem, suggesting that sucrose could play a signalling role.

Several Arabidopsis mutants affected in starch synthesis or mobilisation are late-flowering in non-inductive SD, but not in inductive LD and are still sensitive to photoperiod. This suggests that sucrose is required for flowering in SD. In a starchless mutant, a reduction of the flowering response in inductive conditions correlates with a smaller increase in sucrose level in leaf exudates. The late flowering phenotype of another starch-deficient mutant can be rescued by addition of sucrose to the growth medium.

Sucrose can be supplied to the aerial part of the plants by growing them on vertical plates containing sucrose. In these conditions, flowering of otherwise late-flowering Arabidopsis ecotypes is accelerated, both in LD and in darkness, indicating that sucrose affects flowering not only under SD conditions. Sugar application also promotes flowering of S. alba, although high sucrose concentrations inhibit flowering of several species, including Arabidopsis. On vertical sucrose plates, the late-flowering phenotype of several Arabidopsis mutants is completely or almost completely corrected, whereas ft and fwa mutants flower as late as in the absence of sucrose and GA deficient and insensitive mutants flower later than the wild type.

Sucrose supplied to the roots also accelerates flowering of several late mutants and has no effect on ft and fwa. These results indicate that sucrose requires GAs and the function of FT and FWA to promote flowering. Therefore, sucrose acts either downstream of most flowering-time genes or independently of them and either upstream of FT and FWA or interacting with their products. It has been reported that, in wild-type Arabidopsis plants, high sucrose can slightly delay flowering and reduce FT and SOC1 mRNA abundance, whereas in co mutants accelerates flowering but does not increase FT and SOC1 mRNA levels.

Therefore, the effect of sucrose on flowering time cannot be explained through transcriptional regulation of FT and SOC1. Interestingly, the promotion of flowering by sucrose correlates with LFY upregulation, and treatment with GAs has a synergistic effect with sucrose on LFY expression. All these results underline the complex interactions that exist among putative components of the long-distance signals. Sucrose strongly induces tuberisation, although most observations have been done in in vitro tuberisation assays and cannot be extrapolated to whole soil-grown plants.

Furthermore, tuber growth implies substantial formation of starch, making difficult a distinction between a signalling and a metabolic role for sucrose. Although the rate of tuber growth is affected in potato plants with reduced levels of the sucrose transporter SUT1, which is essential for long-distance transport of sucrose, it has not been reported whether tuberisation time is also altered.

### *Proteins and Peptides*

Intercellular movement of several proteins is relevant for developmental processes, like endodermis specification in the root and flower development. In these cases, cell-to-cell traffic, rather than phloem translocation of proteins, seems to be the operating mechanism. Nevertheless, polypeptides and proteins are involved in long-distance signalling in plants. More than 100 peptides and small proteins have been detected in the phloem sap of Perilla ocymoides and Lupinus albus.

Four small proteins were present in phloem exudates of flowering P. ocymoides and absent in non-induced plants. These peptides are similar to a serine/threonine protein kinase, two purine permeases and ubiquitin and they were also detected in flowering L. albus. It could be tested if any of these peptides plays a role in regulating flowering time. However, the samples used for this analysis were harvested three weeks after the beginning of the inductive treatment. Therefore, the peptides identified may be several steps downstream of the floral stimulus.

## Developmental Processes of Long-distance Signals

### Development and Growth of Shoot

Alterations in the roots can lead to changes in shoot development and growth, suggesting that the roots send signals to the aboveground parts of the plant. Flooding of the soil induces developmental and physiological changes such as leaf epinastic curvature, decelerated leaf expansion and stomatal closure. An increased export of the immediate precursor of ethylene from the roots is likely to be part of the mobile signal that causes these modifications in the shoot.

ABA has also been implicated in this signalling. Recent work points to a role of a carotenoid-derived substance in root to shoot communication. The bypass1 (bps1) mutant of Arabidopsis produces a root-derived signal that arrests shoot development. Fluridone, which blocks carotenoid biosynthesis, partially suppressed the bps1 phenotype, while two mutations that reduce ABA levels, aba1 and aba2, enhanced it. This suggests that the mobile signal produced by bps1 is probably a novel carotenoid, but it is not ABA, the best characterised carotenoid-derived hormone.

Since aba1 mutants accumulate zeaxanthin, the bps1 graft-transmissible signal could be a carotenoid derived from zeaxanthin. The regulation of shoot branching by several plant hormones has been recently reviewed. Auxin transported from the shoot apex in a basipetal direction limits branching by inhibiting axillary meristem outgrowth. Since auxin does not act directly in the bud, it must act through mobile signals from a distance.

Recent work indicates that auxin can act in the vasculature to affect axillary bud growth. Cytokinin has the opposite effect to auxin: it promotes branching. Decapitation-induced branching correlates with increased concentration of cytokinins in the xylem. Application of auxin to the shoot of decapitated plants nearly eliminated the increase in cytokinins, suggesting that the induction of branching is due to a fall in auxin levels that in turn causes an elevation in cytokinins. The interpretation of these results is that auxin reduces cytokinin concentration in intact plants, thus limiting axillary meristem growth.

New evidence shows that cytokinin is not the only signalling molecule acting downstream of auxin. Grafting studies using pea and Arabidopsis mutants with increased branching indicate that the corresponding genes control a long-distance signal that can move from the root to the shoot. These mutants do not show an increase in cytokinin or a decrease in auxin levels. Two of these genes, MORE AXILLARY BRANCHING 4 (MAX4) /RAMOSUS 1 (RMS1) and MAX3, encode members of the carotenoid cleavage dioxygenase family and MAX3 can cleave several carotenoids when it is expressed in E. coli strains that accumulate carotenoids.

Furthermore, RMS1 expression is upregulated by auxin. These data strongly suggest that a novel graft-transmissible carotenoid-derived molecule mediates the action of auxin on shoot branching. The implication of graft-transmissible carotenoid-derived molecules in shoot development and branching indicates that carotenoids can be part of mobile signals. Whether these molecules can be relevant for the regulation of flowering or tuberisation is not known, because the long-distance signals for flowering and tuberisation are transported in the phloem, whereas the signals for shoot development and branching move through the xylem.

### *Development of Leaf*

A new perspective in long-range signalling has been opened by the study of a tomato leaf morphology mutant caused by a naturally occurring gene fusion. When wild-type scions were grafted onto mutant stocks, the new leaves formed in the scion showed the mutant phenotype. More importantly, the mutant mRNA was detected in the scion, indicating that a grafttransmissible mRNA is able to affect a developmental process. Previously, numerous mRNAs had been identified in the phloem sap of pumpkin (Cucurbita maxima). Altogether, this suggests that mRNAs may act as transmissible signals in other developmental processes.

### *Systemic Acquired Resistance*

In response to pathogen attacks, plants produce a long-distance signal that induces disease resistance in uninfected tissues. Salicylic acid is required for the induction of this systemic acquired resistance. Although this compound is known to move in the phloem and for some time had been considered a putative systemic signal in this process, it is more likely that regulates the synthesis of other mobile molecules. Salicylic acid can induce flowering in duckweeds, although most plants do not respond to this compound.

Very recent work shows that salicylic acid is involved in the promotion of flowering in Arabidopsis. Treatment of wild-type plants with salicylic acid accelerates flowering and mutants defective in salicylic acid flower later than wildtype plants, mainly under SD conditions. The acceleration of flowering by salicylic acid is likely to be mediated at least in part by increased levels of FT mRNA. This suggests that flowering and systemic acquired resistance may share mechanisms for long-distance signalling.

The production or transmission of the systemic signal for acquired resistance also requires a gene encoding a protein with homology to lipid transfer proteins. It has been proposed that this protein might help the transmissible signal to enter the vascular system or to move through the plant, suggesting that this signal might be a lipid molecule. Could FT and TFL, which are similar to lipid-binding proteins, play an analogous role in floral signalling?

#### *Responses of Wounded Plants*

In wounded plants, a mechanism similar to systemic acquired resistance is triggered. Wounded leaves of tomato and potato plants release a signal that moves in the phloem to activate defence genes in distant unwounded leaves. Systemin, an 18 amino acid peptide processed from a larger protein called prosystemin, analogous to polypeptide hormones in animals, is essential for long-distance signalling in this defence response.

FT and TFL share sequence similarity with the precursor of an eleven amino acid peptide, hippocampal cholinergic neurostimulating peptide, which affects the development of the hippocampus in animals. This peptide is generated by proteolytic cleavage at the amino terminal end of the precursor. It might be worthwhile testing whether FT and TFL could be proteolytically processed in a similar way to generate peptides that might act in signal transmission.

In the wound response, systemin leads to activation of jasmonic acid synthesis. Recent results obtained with mutants unable to perceive systemin or jasmonic acid and mutants defective in jasmonic acid synthesis suggest that jasmonic acid could be the systemic signal. Jasmonic acid and a structurally related compound, called tuberonic acid, can induce tuberisation in vitro. Application of jasmonic acid to excised stolon tips from potato can induce morphological changes resembling the initial stages of tuber formation. The role of jasmonic acid in tuber induction in soil-grown plants is not clear, but it could be another candidate for the mobile signal.

### *RNA*

Two types of RNA have been found in the phloem of several plant species.

First, different mRNAs have been isolated from phloem sap samples of pumpkin and at least two of them can move across a graft junction from a pumpkin stock into a cucumber scion. Furthermore, as already mentioned, a mutant mRNA affecting leaf development in tomato confers the mutant phenotype in a wild-type scion after grafting onto a mutant stock. Taken together, these results indicate that systemic movement of mRNA can be a mechanism for long-range signalling in plants. This raises the possibility that mRNA could be part of the mobile signal for developmental events such as flowering and tuberisation.

Second, more recently, small RNAs have been detected in the phloem sap of a variety of plants. Small RNAs regulate gene expression through several mechanisms and include microRNAs (miRNAs) and short interfering RNAs (siRNAs). It has been suggested that siRNAs might participate in the systemic spread of RNA silencing, although whether these RNA molecules are the mobile silencing signal or not is a matter of debate.

Endogenous miRNAs have been found very recently in the phloem sap of healthy pumpkin and siRNAs in the sap of viralinfected plants and spontaneously silencing transgenic plants. Grafting experiments have shown that the siRNAs of spontaneously silencing transgenic stocks are grafttransmissible and can induce RNA silencing in non-silencing transgenic scions, strongly suggesting that siRNAs are components of the systemic silencing signal. Accumulation of miRNAs in Arabidopsis requires DICERLIKE1 (DCL1), a homologue of the ribonuclease responsible for processing miRNA precursors in animals.

A mutation in DCL1 affects flowering time, suggesting that small RNAs might affect the induction of flowering. This has been confirmed recently by several reports showing that miRNAs are involved in flowering. A precursor of one of these miRNAs is up-regulated after floral induction and this up-regulation requires the function of CO and FT. Another plant developmental process, leaf morphogenesis, is also under the influence of miRNAs.

Interestingly, a miRNA involved in specification of leaf polarity, miR166, accumulates in the phloem, suggesting that it might be a phloem-mobile signal. However, since miR166 target genes are also involved in vascular patterning, it is unclear whether miR166 is present in the phloem to affect vascular development locally, to move to distant parts of the plant, or both. The hypothesis that miRNAs might have a non-cell-autonomous role in the regulation of flowering time has been formulated recently.

Small RNAs have two features that make them good candidates for the mobile signals: they carry information in their sequence, conferring them high specificity, and their small size makes them easily exportable.

Furthermore, they can induce rapid turnover of mRNAs, allowing fine tuning of responses to environmental and endogenous signals, which is one of the characteristics of the floral stimulus. Testing whether any of the small RNAs or mRNAs present in phloem sap affects flowering time or tuber induction could help to elucidate a possible role of these molecules in longdistance signalling for these processes. An analysis of phloem sap for the presence of candidate mRNAs and miRNAs that regulate flowering time would also contribute to clarify this issue.

# 12

# Hairy Root Cultures of Artemisia annua L.

Secondary metabolites produced exclusive of the primary metabolites essential to sustain the life of an organism. These products have been shown to be very useful for both the plants as well as the animals that synthesise them. The most abundant and structurally dissimilar group of secondary metabolites is the terpenoid family (or isoprenoids). Terpenoids are a class of secondary metabolites, derived from isopentenyl diphosphate (IPP), with nearly 22,000 members exhibiting more than 300 ring systems.

Terpenoids, including the retinoids, the geranylgeranyl and farnesyl protein anchors, vitamins A, D, and E, coenzyme Q, cholesterol, and the steroid hormones have been known to play decisive roles in organisms. Plants direct growth and development via regulatory terpenes that include the gibberellins, the brassinosteroids, and abscisic acid. Some of these compounds are medicinally important, for instance, ginkgolide, which has been shown to delay the inception and progression of Alzheimer's disease as well as taxol, used as an anticancer agent.

Production of certain protective terpenoids, by various plant species, has also been shown to impair the biological processes in herbivores. Numerous plant terpenoids have been found to perform secondary functions, primarily in defence against insect herbivores. Certainly terpenoid-based resins, produced by conifers, have long been studied for their industrial importance and role in defence against herbivores and pathogens.

Since the 1950's, it was known that the central components for terpenoid biosynthesis, isopentenyl diphosphate (IPP) and dimethylallyl diphosphate (DMAPP), are synthesised from activated acetic acid via the mevalonic acid pathway. Today there is a wealth of evidence that shows that an additional pathway exists for the formation of IPP and subsequent terpenoid synthesis in higher plants. The mevalonate pathway is located in the cytosol/ endoplasmic reticulum, and the 2-C-methyl-erythritol-4-phosphate pathway (mevalonic acidautonomous or MEP), which proceeds via 1-deoxyxylulose-5-phosphate, is localised in plastids.

**Artemisinin**

Structurally, sesquiterpenes can be acyclic (farnesol), monocyclic (abscisic acid), or polycyclic (artemisinin). Many also exhibit biological activity, for example, the plant growth regulator abscisic acid (ABA). Artemisinin is now commercially available as a therapeutic against drugresistant strains of Plasmodium, the malarial parasite. Unfortunately artemisinin production by the A. annua plant is usually in the range of 0.01% to 0.4%, but some clones can produce over 1% of the plant's dry weight. Acton et al., reported that artemisinin can also be obtained from artemisinic acid, which can be found at concentrations as much as 10-fold higher than artemisinin in hairy root tissues, and the native plant.

In *Artemisia annua*, the maximum artemisinin accumulation, from field grown crops, takes place at or near the onset of flowering in most strains, hence only one crop can be harvested annually. Also, the highest concentrations obtained at harvest are 0.8% and have been reported to be as low as 0.001% of the plant's dry weight. This is equivalent to 8 mg/g and 0.01mg/g dry weight, respectively. Klayman reported that in clinical trials, one course of treatment requires 2-3 grams of the drug and that repeated recrudescence may call for multiple courses of treatment per year.

Since malaria threatens roughly 40 % of the world's population causing death, morbidity, and significant socio-economic loss, a considerable amount of artemisinin is required. The limited availability from large-scale cropping of A. annua plants coupled with high demand for the therapeutic has spurred scientists to explore alternate methods of production. Chemical synthesis of artemisinin has been attempted. However, due to the presence of the endoperoxide moiety in the structure of artemisinin, present methods for chemical synthesis are expensive and have met with limited success.

An alternative to field-grown crops or to chemical synthesis has been large scale in vitro plant tissue culture. In vitro cultures may potentially constitute useful and easily manipulated systems for producing valuable biologically active compound in plants that do not require labor-intensive methods. However, in vitro production systems are still not cost effective and produce low yields.

## In Vitro Methods for Overproduction of Secondary Metabolites

### *Undifferentiated and Differentiated Culture*

Alternative methods such as cell suspension cultures

have been sought to deal with problems of low concentrations in whole plants of secondary metabolites, like artemisinin and. While suspension cultures can be transformed with novel genes and are relatively easy to scale-up, they are not very genetically stable. Disappointing yields coupled with instability have required new methods for improving production of secondary products. Toward this end, genetic manipulation of dicotyledonous whole plants using modified Ri or Ti DNA segments from agropine- and mannopine-type strains of Agrobacterium, has resulted in neoplastic roots that have augmented growth and secondary metabolite production rates.

Secondary product levels in *in vitro* differentiated tissues, are frequently either comparable or greater than those found in either roots or shoots of the intact plant. These compounds are also produced concomitantly with growth. Therefore, continuous extraction of secondary metabolites from actively growing hairy roots is possible. Moreover, the transformants are genetically stable over a long period of time with consistent biosynthetic capabilities. Also, hairy roots do not require the application of growth regulators.

In 1994, Weathers et al. reported high levels (0.4% of the dry weight) of artemisinin in hairy root cultures of *A. annua*. While one cannot preclude the value of suspension cultures, their inherent genetic instability dilutes their potential as feasible secondary metabolite production systems. Conversely, the long term genetic stability, and consistent biosynthetic capabilities of transformed hairy roots predisposes these differentiated cultures as a better means for secondary metabolite production *in vitro*.

## Effects of Abiotic Factors

Many factors including light, medium composition, pH and most notably, the exogenous application of plant growth regulators affect secondary metabolism in in vitro hairy root cultures. The relative impact of these influences is usually species dependent. Researchers studying alkaloid production in transformed hairy root cultures noted that optimisation of the medium, specifically mineral composition and sucrose concentration can play either a neutral or positive role in growth of the roots as well as secondary metabolite production.

The external pH of the medium changes over the course of batch growth, and has also been shown to influence the release of secondary products from root cultures. Endogenously produced in a variety of plant tissues, plant hormones are present in trace amounts and are responsible for an array of developmental processes that involve abiotic as well as biotic factors.

Since the discovery of these signalling compounds in the mid 1920's, a considerable amount of research has been performed examining the influences of both endogenous and exogenous concentrations of phytohormones on plant cell physiology. Currently, we know that these growth regulators direct culture growth, influence cell secondary metabolite accumulation, and promote morphological diversity in a concentration dependent manner. There are five main phytohormones: gibberellins, auxins, cytokinins, ethylene, and abscisic acid. The effects of this hormones on the hairy root culture are described below.

### *Effects of Gibberellic Acid*

Fundamental to hairy root culture systems, is their ability to grow without the addition of plant growth regulators.

Interestingly, the addition of an individual growth regulator, like gibberellic acid, has yielded definitive evidence that GA can play an important physiological role in root morphology and secondary product accumulation. In studies of the effects of GA on hairy root morphology, Bais et al. reported that low concentrations (0.5 mg/l) of GA increased the biomass accumulation in the hairy roots of *Cichorum intybus*, due to high levels of root elongation, lateral root branching, and primary root growth almost 2-fold over the hormone-free controls.

Vanhala et al. observed, in the presence of GA no loss of *Hyocyamus muticus* root integrity but that the hairy root cultures seemed to age rapidly compared to roots not supplied with GA. Surprisingly, this premature aging phenomenon did not decrease biomass accumulation. Ohkawa and coworkers observed that GA stimulated elongation and lateral branching of the hairy roots of *Datura innoxia*. Their roots were transformed with TL-DNA (the portion of Ti DNA lacking the IAA synthesising genes) and displayed rapid growth and copious amounts of lateral roots in the presence of GA at concentrations between 0.01-1.0 mg per liter.

It was concluded that GA, rather than auxin, was responsible for controlling root morphology in *D. innoxia*. Also, exogenously supplied GA3 enhanced branching in Brugmansia candida roots. A pronounced influence on growth was observed by Liu et al. in the hairy roots of *A. annua* in the presence of GA at a concentration of 5 mg/l. Of all the hormones studied in that report, GA was the only one to yield high levels of biomass. This may be a reflection of the increase in root elongation and lateral branching as seen in other studies involving hairy root cultures.

For example, in laboratory Smith et al. observed that GA3 concentrations of 1-10 ng per liter provided the most significant increase in biomass during exponential growth in *A. annua* hairy roots. In terms of secondary metabolite production, variable results have been obtained using GA, and are usually species specific. For example, Vanhala et al. observed that in *H. muticus* hairy roots, GA decreased the accumulation of the secondary metabolite hyoscyamine, compared to roots cultivated in hormone-free medium.

Conversely, Liu et al and Bais et al observed that the addition of GA to the growth medium increased artemisinin and coumarin content in the hairy roots of *A. annua* and *C. intybus* L. CV. Lucknow local, respectively. Since GA is known to induce flowering in some plants, and maximum artemisinin accumulation in field-grown crops occurs at or near the onset of flowering in most strains, this may explain the increase in artemisinin. In the case of coumarin production, Bais et al. observed that coumarin accumulation was strictly related to growth, suggesting that GA plays a key role in the production to coumarin.

### *Effects of Auxin*

Hairy root auxin sensitivity has received great attention. Moreover, studies on the biochemical relationship between exogenous and endogenous auxin levels using these differentiated tissues have returned interesting findings in the area of valuable root-derived biologically active compounds. By and large, researchers investigating the physiological role of exogenously applied auxin in root growth and secondary metabolite production have established that these signal molecules can great impact plant tissue stability and secondary product accumulation either individually or through phytohormone-phytohormone interactions.

Sauerwein et al. reported on the exogenous phytohormone-phytohormone relationship displayed in cultures of *Hyocyamus albus* and described the formation of pure calli in the presence of both exogenous auxin and cytokinin. They observed that the single addition of only IAA to the medium resulted in partial disorganisation of the hairy root morphology. However, growth rates increased, upon the transformation strain used, when there were low levels of auxin and cytokinin (1:1mg/l ratio).

Rhodes et al. also observed disorganisation of the hairy root matrix of *Nicotiana rustica* when cultures were supplied with auxins and cytokinins. However, these effects were minimised by increasing the root inoculum size. Bais et al. also showed that when high levels of exogenously applied auxins, specifically IAA and NAA, were added in the presence of low exogenous cytokinin levels, there was a decrease in biomass in the hairy root cultures of *C. intybus*, unless the root inoculum size was increased. Although others have shown that auxin addition to a variety of hairy root cultures stimulates growth, the response is likely species specific.

For example, Lin et al. demonstrated that auxin had influence on hairy root cultures of Linum flavum. Taken together, these studies on the effects of auxins on hairy root growth show that, although there is some species variability, the ratio of the exogenously added hormone to root biomass is critical to achieving good root growth.

In *Tagetes patula* hairy roots, Arroo et al. showed that the addition of IAA inhibited secondary metabolite accumulation. In contrast, addition of either IBA or NAA stimulated ajmalicine and ajmaline production over the

*Rauwolfia micrantha* hairy root cultures in hormone-free medium, whereas Rh et al. observed a decrease in nicotine content in the hairy roots of Nicotiana rustica when roots were supplied with auxins plus cytokinins. Luczkiewicz et al. discovered that the production of the sesquiterpene lactone pulchelin E was enhanced in hairy roots of *Rudbeckia hirta* compared to that of callus and suspensions cultures in the presence of auxin.

#### *Effects of Cytokinin*

Cytokinins have been shown to direct shoot growth in tissue culture. The intrinsic ability of cytokinins to alter tissue morphogenesis favouring the production of shoots may be the reason it has routinely received such little attention in hairy root cultures. In any case, exogenous cytokinins impact growth and secondary metabolite accumulation in a phytohormone-to-phytohormone ratio-dependent manner that can be toxic and even lethal. Likewise, they can induce spontaneous shoot formation in hairy root cultures when administered individually at high concentrations.

Indeed, low cytokinin to auxin levels have been shown to induce rapid disorganisation in hairy root cultures of *C.intybus* as well as decrease root growth and the ability of root cultures of *C. intybu* L. cv. lucknow local, and B. candida to produce secondary products. Vanhala et al. similarly demonstrated that the root growth rate and secondary product accumulation did not change despite addition of exogenous cytokinins to *H. muticus* cultures.

In contrast, Sauerwein et al. observed that when the ratio of exogenous cytokinin to exogenous auxin was 1:1, both growth and alkaloid accumulation were enhanced in hairy root cultures of *H. albus*. In *A. annua*, addition of cytokinin (BAP) to the medium of hairy roots resulted in the disorganisation of the root matrix with little effect on production of artemisinic compounds.

### *Effects of Ethylene*

Literature reporting on the affects of exogenous ethylene on hairy root cultures is virtually non-existent. Perhaps this is due to the fact that ethylene alters the in vitro behaviour of the plant tissues and promotes senescence. Ethylene can promote or inhibit, in a species dependent fashion, the production of secondary products. Shibli et al. reported that increasing ethylene content in the medium caused a reduction in pigment production and callus growth in *Vaccinium pahalae*. Fulzele et al. concluded that exogenous addition of ethylene to the culture medium stimulated terpenoid synthesis in *A. annua* L. plantlet cultures grown in a bioreactor.

### *Effects of Abscisic Acid*

Exogenous application of ABA to cell suspension cultures appears to stimulate secondary product accumulation, while inhibiting biomass yield in a time dependent manner. For example, the addition of 5 mg of abscisic acid per liter after 12 days growth of *T. chinensus* suspension cultures boosts paclitaxel accumulation roughly 5 times that of the controls. They also observed decreases in biomass in all cultures where ABA was added on day 0. Conversely, cultures where ABA was added at day 12 displayed minimal affects on biomass. On the other hand, Vanhala et al. demonstrated that exogenous application of ABA strongly inhibited hyocyamine accumulation in hairy root cultures of *H. muticus*. They did not observe any adverse effect on root biomass.

## Combinatorial Phytohormone Treatment

The influences of these growth regulatory compound are numerous and far reaching. Studied individually, these hormones have increased our understanding of factors that affect growth and secondary metabolism of in vitro plant cultures. The information gained from the experiments above will inevitably guide the field of metabolic engineering in the future. However, one immediate objective, thus far overlooked by others, is the employ of hormone combinations in order to increase biomass and secondary metabolite production in hairy root cultures. Auxin:cytokinin combinations have met with success in a limited number of plant species. Since hormones regulate a multiplicity of plant functions, it is reasonable to assume that combination of hormones at different concentrations may also affect terpenoid biosynthesis.

### *Multi-stage Culture Process*

It was observed some time ago that many secondary metabolites are often produced after most growth has ceased, or at least not during exponential phase. Indeed the first commercially produced plant product, shikonin, is produced during a second stage, in a new medium, and after growth has mainly ceased. This process is known as a two-stage culture and in some cases, growth and secondary metabolite production has been as high as five times that of a one-staged culture that used only an original medium formulation.

# 13

# Diagnosing Problems Due to Plant Growth Regulators

Plant growth regulators (PGRs) are chemicals that are designed to affect plant growth and/or development. They are applied for specific purposes to affect specific plant responses. Although there is much scientific information on using PGRs in the greenhouse, it is not an exact science. Achieving the best results with PGRs is a combination of art and science—science tempered with a lot of trial and error and a good understanding of plant growth and development.

## Growth Regulator Herbicide Symptoms in Soybean

Leaf cupping and malformation in soybean are not well understood; however, it seems that there are at least three circumstances that may lead to these symptoms:

– soybean plants are exposed to a plant PGR herbicide (e.g., 2,4-D and dicamba);

– application of a postemergence herbicide to soybean induces the response, and

– environmental conditions stimulate the symptoms.

In some situations, it is easy to determine the cause of injury, whereas in others it may be impossible to conclusively determine why leaf cupping has developed. The first scenario may occur due to drift of the growth regulator herbicide from an adjacent area or may be due to use of a contaminated sprayer to apply herbicides directly to the soybean. Examples involving off-target herbicide movement, whether it is caused by spray drift or volatilisation, show a pattern. If there is no pattern evident, herbicide drift is not likely to be involved.

Symptoms frequently develop within a week of a postemergence herbicide application to soybean. Sometimes the entire field is affected, whereas in other fields the damage is limited to specific sprayer loads. In these situations, it can be difficult to determine whether the sprayer contained residues of a PGR herbicide or whether the problem was induced by other factors. Concentrations of a PGR herbicide capable of inducing symptoms may remain in the sprayer even after thorough cleaning procedures or after several loads of non-PGR have been run through the sprayer.

However, we have seen numerous examples where leaf cupping developed after application with a sprayer dedicated to spraying soybean, thus minimising the possibility of sprayer contamination. Leaf cupping has been observed after applications of all types of herbicides; thus, the response does not seem to be related to the mode of action of the herbicide. Roundup Ready soybean plants seem to be as likely to show the response as traditional varieties.

#### Leaf Malformations

Some fields have developed leaf cupping in the absence of herbicide applications, most commonly during conditions of rapid growth. It is possible that heat and moisture stress induce leaf cupping and distortion. Under stressful conditions, the balance of naturally occurring hormones in the plant apparently is disrupted, resulting in symptoms characteristic of PGR herbicide damage. When this situation develops, the entire field will probably demonstrate symptoms and there will not be a drift pattern. When PGR herbicides are not involved in the occurrence of the injury symptoms, soybean typically resumes normal growth shortly after the cupped leaves are observed.

Frequently, two or three leaves develop symptoms and then normal growth resumes. Soybean yield should not be impacted under these situations. The potential for a yield response is greater when a PGR herbicide is involved; however, it is impossible to determine the extent of yield loss by examining symptoms that develop after the exposure. Research has shown that low levels of injury early in the season from PGR herbicides usually do not impact yields. The only reliable method of determining a yield response is comparing the yield of the injured soybean with an area of the same field that is unaffected by the herbicide. Usually, a valid comparison is not available to help determine the impact of the herbicide damage on yield.

Over the last several years unexplainable soybean leaf cupping has occurred widely throughout the Midwest. We suspect that environmental conditions that cause rapid soybean growth, often preceded by stress conditions, are instrumental in the occurrence of the symptoms. Stress from herbicide applications also may induce this response. No one adjuvant, crop oil, seed oil, or surfactant seems to be involved. In these situations, the development of deformed leaves seems to be temporary and new leaves have not been affected, thus suggesting there is no impact on soybean yield.

## Plant Growth Regulators on Floricultural Crops

Plant growth regulators (PGRs) are excellent tools to help control undesirable stretch of floriculture crops. When used at the correct stage of development and rate, they provide excellent height control. Misapplications can lead to catostrophic results, which lower plant quality.

For best results, PGRs should be handled as production tools, like water and fertiliser. They should not be used as crutches for poor management of other cultural practices. PGRs should be an integrated part of the crop production cycle. They are most effective when applied at the appropriate times to regulate plant growth or development. In other words, growth retardants cannot "shrink" an overgrown plant. They must be applied before the plant is overgrown to avoid plant stretch. Plant growth regulators are classified as pesticides.

### *Height Controlling Growth Regulators*

Most of the PGRs used in the greenhouse are used to regulate shoot growth of bedding plants, garden mums, poinsettias, and other containerised crops. These PGRs are referred to as "growth retardants." Typical growth retardants are B-Nine, Cycocel, A-Rest, Bonzi, and Sumagic. These PGRs reduce plant height by inhibiting the production of gibberellins, the primary plant hormones responsible for cell elongation. Therefore, their effects are primarily on stem, petiole and flower stalk tissues. Lesser effects are seen in reductions of leaf expansion, resulting in thicker leaves with darker green colour. Other benefits of using these PGRs in plant production include improved plant appearance by maintaining plant size and shape in proportion with the pot.

Plant growth retardants also increase the stress tolerance of plants during shipping and handling and retail marketing, thereby improving shelf-life and extending plant marketability. Remember, growth retardants do not reduce plant size. They reduce the plant's growth rate. You must apply the growth retardant prior to the "stretch." Look for recommendations on the PGR label for time of application. These recommendations will be given in terms of plant development or plant size as opposed to production time. For example, the Sumagic label specifies that pansies should have attained a minimum height of four inches prior to application. The Bonzi label says that bedding plant plugs should be treated at the one to two true leaf stage and bedding plants (after transplanting) at two inches of new growth or when the plants reach marketable size.

Generally, growth retarding PGRs should be applied just prior to rapid shoot growth. This is usually one to two weeks after transplanting a plug, after the roots are established and as the plant resumes active growth; on pinched plants, it is after the new shoots are visible, just starting to elongate. This is where the art of plant growth regulation is most important. You must learn how your crop grows and when to intervene to obtain the desired results.

Many growers use multiple applications of growth retardants to better control plant growth. A single application at a high rate early in the plant production cycle may be excessive if growing conditions are not as good as expected. An early application at a lower rate provides more flexibility, but the tradeoff is in the additional labour involved with a second application if it becomes necessary. Some growers improve crop uniformity by using multiple applications of lower rates to affect small corrections in plant growth.

Be aware that excessive rates of Bonzi or Sumagic can cause extremely persistent growth reductions in the flat or even in the landscape. It is always a good idea to evaluate the long-term effects of your treatments by growing some out for yourself and talking with your customers. Be careful to avoid very late applications, especially of Bonzi or Sumagic as they may delay flower opening on bedding plants. However, drench applications of Bonzi have provided excellent control of poinsettia height very late in the production cycle without causing the reduction in bract size accompanying late spray applications. Learn the art of using PGRs for plant growth regulation.

#### *Chemical Pinchers*

Another group of PGRs used in floricultural crops are those that enhance branching, including Florel, Atrimmec and Off-Shoot-O. These PGRs are frequently called "chemical pinchers" because they generally inhibit the growth of the terminal shoots or enhance the growth of lateral buds, thereby increasing the development of lateral branches. They can be used to replace mechanical pinching of many crops. Often this increased branching also will reduce the overall height of the plant. The ethylene released inside the plant by Florel also inhibits internode elongation, keeping treated plants more compact than untreated plants.

Florel also affects flowering. If you are looking for enhanced branching, you must have sufficient growth on the plant to allow for sites of lateral development. They cannot enhance lateral branching if there are no laterals on the plant. Again, read the label for details of when to apply for optimum response. You may need to consider combinations of PGRs. For example, if you apply Florel to enhance the branch development of 'Wave' petunias in a hanging basket, you will probably need to follow up with a treatment of a plant growth retardant like Bonzi to control the elongation of those new laterals. Always consider the side effects of treatments.

### *Flowering Growth Regulators*

Plant growth regulators can be used to enhance flowering (GibGro) or to remove flowers (Florel). To improve flowering, GibGro, which contains the growth promoter gibberellic acid, can be used to substitute for all or part of the chilling requirement of some woody ornamentals typically forced in the greenhouse, including azalea. [A broad use label was submitted for EPA approval in 2001 for Pro-Gibb (Valent USA) which would include camellia, hydrangea, and a variety of other floricultural crops.] Special attention must be given to the stage of flower bud development for successful treatment. In addition to overcoming dormancy, these compounds can improve flowering and/or bloom size of camellia, geranium, cyclamen, spathiphyllum, statice, and calla lily.

Again, timing is critical since late applications, or excessive rates, may cause excessive plant stretching resulting in weak, spindly stems. Cycocel used to control stem height of hibiscus and geranium also improves early flowering. Flower removal is especially desirable for stock plants maintained for cuttings of vegetatively propagated ornamentals, like geraniums, fuchsia, begonias, or lantana. Florel (ethephon) is the primary compound used for flower removal. Once ethephon is absorbed by the plant it is converted to gaseous ethylene, a natural plant hormone effective in many plant processes. Ethylene is the primary hormone responsible for flower senescence and fruit ripening.

It is the "postharvest" hormone. With proper rates and timing, it will remove unwanted flowers from stock plants or from plugs or young bedding plants. Flower removal diverts more energy into vegetative growth, increasing the number of laterals available for cuttings on stock plants, and promoting increased branching of plugs and finished plants, which increases fullness in hanging baskets or other containers. Early flower removal also allows synchronisation of flowering of a container for a more dramatic appearance or for flowering on a specific marketing date. Since initiation and development of flowers requires time, Florel should not be used on crops within six to eight weeks of marketing.

### Symptoms of PGR Overdose

Common problems include phytotoxicity, delayed flowering, and stunted growth.

#### *Phytotoxicity*

Phytotoxic means harmful or lethal to plants. Phytotoxicity is the degree to which a chemical or other compound is toxic to plants. All types of pesticides can injure or kill plants. Herbicides are especially hazardous to plants because they are designed to kill or suppress plants. Some insecticides and fungicides can also harm plants. Phytotoxic effects caused by herbicides can be from spray droplets, soil residues or vapours contacting sensitive plants. Plants can also be harmed by herbicides which move off target in water or soil or when sensitive crops are planted in fields too soon after a herbicide treatment was applied for a previous crop.

Phytotoxic properties of pesticides are usually associated with specific formulations (wettable powder, emulsifiable concentrate, granule, etc) or specific plants rather than groups of pesticides or plants. A pesticide label may indicate whether the pesticide could be phytotoxic, and may list plants or varieties that are sensitive. Phytotoxic effects can range from slight burning or browning of leaves to death of the plant. Sometimes the damage appears as distorted leaves, fruit, flowers or stems. Damage symptoms vary with the pesticide and the type of plant that has been affected.

*Glyphosate damage to apple tree*

*Glyphosate drift damage to cucumber*

*Dimethoate injury to cherry tree*

*Kresoxim-methyl drift injury to 'Sweetheart' cherry*

Phytotoxicity is not necessarily caused by the active ingredient. Plant damage can also be caused by: the solvents in a formulation, impurities in spray water, using more pesticide than listed on the label, or poorly mixing the spray solutions. Condition of the plant at the time of treatment can affect phytotoxicity; stressed plants may be more susceptible. Environmental conditions such the temperature, humidity, and light can influence phytotoxicity.

High temperatures can speed up pesticide degradation and volatilisation, but may also result in increased phytotoxicity for some products. UV light rapidly breaks down many pesticides. Soil properties such as texture, temperature, moisture, microbial activity and pH also influence phytotoxicity. Higher pH soils are less binding and may increase photoxicity. High microbial activity can reduce phytotoxicity. Minimise the risk of phytotoxic effects by:

– reading and following label directions, especially the correct rates and timing, and being aware of potential weather effects,
– avoiding application of pesticides when drift is likely to happen,
– waiting for the correct planting times,
– if unsure, conduct a simple field bioassay by treating only a few plants, before the treating whole block to check for phytotoxic effects, especially when growing new culitivars.

Symptoms typical for each PGR are shown.

### *A-Rest*

It is a safe chemical with limited reports of phytotoxicity. Phytotoxicity symptoms appear as necrotic spots on leaves of plugs and poinsettias. Problems are more pronounced with high temperatures (>70°F).

*Phytotoxicity on poinsettias sprayed with 99 ppm A-Rest when temperatures were >70°F.*

### B-Nine

Phytotoxicity is not a problem with B-Nine. Because of its lower degree of activity, multiple applications may be required.

*A late B-Nine application to ranunculus inhibited the elongation of the flower spike.*

These multiple applications can cause flower delay or reduced flower stalk elongation if applied too late.

#### *Cycocel*

Foliar sprays of Cycocel can cause a yellow spotting or discolouratoin of newly expanding leaves. Damage to the chloroplast causes the yellow spotting and symptoms are evident within 3 to 5 days after application.

*Typical leaf yellowing caused by Cycocel phytotoxicity of geraniums.*

In general, problems occur when Cycocel rates exceed 1,500 ppm. Expansion of new leaves generally cover damaged leaves within a few weeks.

#### *Bonzi and Sumagic*

Both Bonzi and Sumagic are triazole type chemicals and have a high degree of activity.

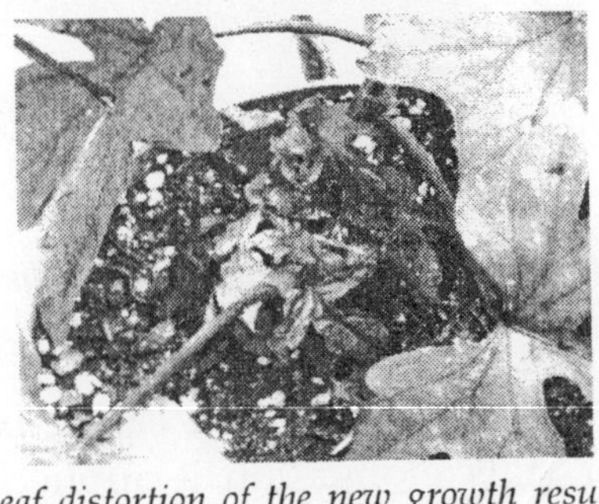

*Severe leaf distortion of the new growth resulted with excessively high rates of Sumagic.*

*Excessive rates of Bonzi caused lower leaf necrosis, stunted growth, and a change in the flower orientation (to upward facing) of pot sunflower.*

Over-application of these chemicals can result in leaf necrosis soon after application, darker green foliage with crinkled leaves, and severely stunted growth. Delay in flowering can also occur.

#### *Florel*

Overdoses of Florel are mainly associated with delayed flowering.

*Uneven spray applications of Florel to an individual plant can result in differences in growth and flowering.*

Applying Florel to plants under stress (temperature or water) also can result in leaf yellowing.

## Applications of Growth Regulators

*Spray Applications*: The pesticide label not only contains information on restrictions but also much information on using the product effectively. The label will identify the target tissue for that PGR—B-Nine is only effective as a foliar spray whereas Bonzi and Sumagic sprays must reach the stems or roots. When making spray applications, look at the physiological development of the plant to see that there is sufficient plant material at the correct stage of growth to make the treatment effective and to accomplish your goal. Generally, there should be sufficient foliage or stems to absorb the PGR.

Uptake and effectiveness of a PGR also depend on selecting the application technique that will ensure proper coverage of the target tissue. B-Nine is not soil active and is fairly mobile in the plant. Therefore, a foliar spray application, wetting most of the foliage, will provide a fairly uniform reduction in growth of sensitive crops. However, the triazoles, Bonzi and Sumagic, are absorbed primarily by stem tissue and then translocated upwards in the plant. Therefore, consistent and complete coverage of the stems is necessary for uniform effects.

In other words, if the stem of one lateral receives an inadequate amount of spray, it will grow faster than the others, resulting in a poorly shaped plant, most noticeable in potted crops like poinsettia or chrysanthemum. The triazoles also are very "soil active" which means they may be adsorbed to particles in the media and become available to the plant through root uptake. Therefore, drenching is a very effective application method for these chemicals in crops where it is economically feasible. The label will provide a recommended application volume for sprays or drenches, especially for chemicals that are soil active.

All foliar applications of PGRs should be applied on an area basis, i.e., uniformly spray the area where the plants are located with the recommended volume of solution. Do not spray individual plants or spray to reach a subjective target like "spray to glistening." Since every applicator will have a slightly different definition of these goals, there will be no way of recommending appropriate rates or obtaining predictable results. For soil active PGRs, dosage equals the concentration of the solution multiplied by the volume applied in the treated area. Therefore, to improve predictability, the label-recommended spray application rates are generally set at 2 qt. finished spray per 100 sq.ft., a comfortable walking pace for applicators with hand-held sprayers.

Since Bonzi and Sumagic are soil active, precautions should be taken to avoid over-application with sprays. Spray applications require more attention to detail, because overspray material lands or drips onto the medium. Remember that dosage equals concentration times volume.

*Spray Equipment.* To assure proper spray volumes, your compressed air sprayer should be equipped with a pressure gauge and regulator and you should consistently use the same nozzle for all PGR applications. Your sprayer should be calibrated by determining the output of the chemical with the selected nozzle at the selected pressure within a specified time period. Spray droplet size also affects response with smaller droplet sizes providing better coverage, but only up to a point. Mist or fog type applicators do not provide adequate volume for coverage of plant stems and have not been effective when used with compounds like Bonzi and Sumagic.

*Applying Drenches.* Drenches have several advantages over sprays. Drenches generally have less effect on flower or bract size and tend to provide longer lasting growth regulation than sprays. Drenches are easier to apply uniformly than sprays because the drench volume is easily measured, and when applied to moist media, it is easy to obtain good distribution of the PGR in the media. Therefore, the resulting growth regulation is frequently more uniform.

The label specifies the recommended volumes for drench applications to different size pots or types of media. In general, 4 fl. oz. of drench solution is applied to a six-inch "azalea" pot, and that volume is adjusted up or down with pot size to obtain a volume where about 10% of the solution runs out the bottom of the pot when the media is moist. Remember that the amount of active ingredient applied to plants using soil-active PGRs is a product of the concentration (ppm) of the solution and the volume applied.

In some cases, drench application recommendations are given in terms of milligrams of active ingredient (mg a.i.) per pot. For Bonzi, the label provides mixing directions for mg a.i. solutions for Bonzi, or, you can use the NC State University "PGR Calculator" to obtain solution directions for drench recommendations using this format.

Other methods of applying PGRs directly to the media have been developed and labelled. For example, Bonzi and A-Rest are labelled for chemigation or application through the irrigation system. These are generally limited to flood (sub-irrigation) or drip irrigation, not overhead sprinkler systems. Again, rates vary with the volumes used and method of application. Bonzi applied once by sub-irrigation requires 50% to 75% of the amount of Bonzi that is applied in a typical drench application. Read and exactly follow the label for chemigation applications.

*Other Types of PGR Applications*: Three other methods of providing a drench type application of soil-active PGRs on a more economical scale are being used by growers.

One is media surface application sprays. These are spray applications made to the surface of the media of filled flats or pots. The treatment is applied at normal to high spray volumes, but since it is applied to the media surface it is activated by irrigation and is available to the plant in the root zone. Both Bonzi and Sumagic are labelled for this method of application. Rates are lower than used for sprays, but higher than used for drench applications.

A second method is called "sprenches" which is a high volume foliar spray that results in runoff into the media, providing a drench effect. Rates are lower than those recommended for sprays.

A third technique is called "watering in" where the PGR, A-Rest and Bonzi are currently labelled, is injected into the irrigation water and applied in each irrigation at very low rates of active ingredient.

All of these application methods use the relationship between rate and volume to provide the desired control and preferred application methods. Again, you must develop techniques that fit your production methods and your growth management preferences.

*Beware of Bark*: For all media applications, be aware that soil-active PGRs tend to be tied up by bark particles in the media which makes it less available to the plants. Therefore, if your media mix is high in bark you will need to compensate for this unavailability by adjusting your application rates up for any type of drench or media applications.

*Growing Conditions*: Look also for label recommendations on time of day or condition of the plant for optimum treatment response. Generally, a healthy, unstressed plant growing under low evaporative conditions, e.g., early in the morning or late in the afternoon, is most responsive to treatment. To maximise uptake, the chemical must remain in contact with the leaf long enough to be absorbed. This time varies for the different PGRs.

Plants treated with B-Nine or Florel should not be overhead irrigated for at least 12 hours after treatment, but plants treated with Bonzi or Sumagic may be irrigated one hour after treatment. Spraying when the treatment will not dry quickly increases absorption of the active ingredients and increases the effectiveness of the treatment. Read the label for any warnings on how irrigation or environmental conditions will affect plant response to the PGR treatment.

# 14

# Plant Growth Regulators and Biotechnology

What is being done with the new technology of biotechnology in the arena of plants. Today, there are products in the field and the marketplace. The new technologies of genetic engineering can result in more environmentally friendly plant pest-protection, foods with enhanced nutrition, more accurate and sensitive diagnostics, foods with improved processing and marketing characteristics, better and more efficient medical delivery, new methods for removing contaminants from soils and waters, and creation of products that are presently being made from nonrenewable resources. Many of these changes might be considered under the broad definition of a "plant growth regulator".

While products of this technology once were confined to the research laboratory, this is no longer the case. These and other new crops represent a substantial percentage of actual production acreage in the U.S. In 1998 50% of the cotton acreage, 30% of the soybean and 20% of the maize acreage in the U.S. were genetically engineered. In the U.S. some 60 million acres of G.E. crops were planted in 1998 and the projections for 1999 are that this number will increase, with a wider variety of products available to farmers and present in the marketplace.

### Pest Control

*Herbicide tolerance*: A major agricultural advance occurred with the identification of selective herbicides that can be used to eliminate weeds without affecting the crop plant. But achieving complete selectivity has been problematic for many crops. An alternative to finding new selective herbicides is to genetically engineer the crop to resist existing herbicides. The engineered mechanisms can include modifications in the site of action of the herbicide, in its detoxification or in the uptake and translocation of the herbicide to its site of action. Such crops represent the first and most frequent application of biotechnology to agriculture.

Of the total of 5900 permits and notifications issued through the end of 1998 by the USDA Animal and Plant Health Inspection Service (APHIS), 27% or 1569 permits were for herbicide-tolerant crops. Based on data from APHIS, work is currently being done on six herbicides: bromoxynil, glyphosate, imadazolinones, phosphinothricin. In many cases, the herbicide tolerance is based on a modified site of action through the use of mutant or altered forms of the target protein.

In these instances, the altered enzyme must be properly compartmentalised (chloroplast, cytoplasm, nucleus) and it must be assembled in a metabolically active form at its target location. This is the case with glyphosate where the herbicide blocks a pathway that leads to the synthesis of several amino acids; blockage of this pathway leads to starvation for amino acids. Genetic engineering strategies have focused mainly on using mutant target enzymes that are resistant to glyphosate. A similar type of strategy is used with another group of herbicides, the imadazolinones and sulfonylureas, which also inhibit a key step in the biosynthesis of a different set of amino acids.

Other approaches to engineering herbicide resistance involve the introduction of genes for enzymes that degrade and detoxify the herbicides. One example in this category is bromoxynil in which resistance was engineered through the introduction of a bacterial gene that codes for an enzyme that converts bromoxynil to an inactive metabolite. Another example is glufosinate. Resistance to this herbicide can be engineered using a bacterial gene that acetylates and inactivates the herbicide. The last example is 2,4-D, a plant hormone that causes arrested development.

Again, genetic engineering strategies employ a bacterial gene that encodes an enzyme that degrades 2,4-D. Cotton has been engineered to be resistant to this herbicide in order to prevent crop loss due to spray drift. Although progress in engineering herbicide-resistant crops has been rapid, there are some problems with this approach. Technical problems include finding an effective strategy for engineering resistance and identifying the target site of the herbicide or a degradative or inactivating factor. In addition the resistance factor must be engineered to be expressed in the proper place and in sufficient levels.

The herbicides for which tolerance is being engineered are generally low-use rate, low-toxicity, rapid-turnover herbicides. Despite this, there are some environmental issues which, although not unique to genetically engineered varieties, must be taken into consideration.

- Generation of herbicide-tolerant weeds due to monoherbicide application. Frequent and long-term application of a single herbicide becomes more likely because of the use of specific herbicide-tolerant varieties. The tendency of the industry to engineer a large number of crop species with tolerance to a single herbicide will create pressure for this to occur
- Transfer of resistance genes to related weeds. Many crops in the U.S. in which tolerance is being developed do not have weedy relatives here; exceptions are sugarbeet, certain cucurbits, radish, canola, oats, and carrots. As seed is moved internationally, however, the possibility of outcrossing with wild relatives in centres of diversity will increase; control measures should be instituted. In general, however, unless pollen travels over long distances to pollinate weeds located far away, this problem will be minimal.
- Encroachment of weeds not affected by the herbicide, i.e. weed shift. Development of elite germplasm resistant to a single herbicide, without comparable varieties resistant to other herbicides with dissimilar modes of action, could lead to an exacerbation of the weed shift problem.

*Insect resistance*: A now standard method for engineering insect resistance in crop species is the use of the toxin genes from Bacillus thuringiensis; this is the basis for Bollgard cotton and the commercially available maize that is European corn borer-resistant. These toxins are activated by cleavage in the midgut of insects; however, making engineered toxins that are effective against certain insects has not been straightforward. For example for a certain insect pest of alfalfa, Spodoptera armyworm, the situation was not straightforward. It required some extensive in vitro testing of synthetic toxins.

In the end a specific synthetic delta-endotoxin gene was identified, which when expressed in optimum amounts in alfalfa, shows evidence of protection against the Egyptian cotton leafworm (*Spodoptera littoralis*) and the beet armyworm (*Spodoptera exigua*). In other cases new types of Bt's are being discovered and also some insecticidal compounds from other microbial species.

*Viral resistance*: Infection with viruses is a problem with many crop species; however the sources of natural resistance are limited and chemical protection is not effective in many instances. A standard method for engineering viral resistance is the use of the coat protein gene. This approach has been used successfully for many viruses in many different plant species and the Freedom II squash was released in the marketplace last year. In the coming year genetically engineered papaya will enter the marketplace. This crop results from efforts to protect it against papaya ringspot virus, which had devastated certain production areas in Hawaii. Because there are limitations to the coat protein approach in terms of uniform applicability and broad protection to related viruses, it is necessary to develop alternative strategies that complement or replace the coat protein strategy.

*Fungal and bacterial resistance*: Methods for protecting against fungal pathogens involve the identification of natural plant protection compounds that, when expressed in higher amounts in appropriate tissues of the host plant, can lessen fungal damage. These are sometimes compounds deriving from secondary metabolic pathways. Work on compounds, such as these, can lead to approaches where increased or broader-spectrum disease resistance can result from modifying expression levels or timing of unmodified secondary metabolite (e.g. phytoalexin) or modifying the phytoalexins themselves to create new phytoalexins with novel activities.

In addition natural defence genes have been isolated for resistance to various pathogens in different plant species. Of the first three such genes that were isolated, it was discovered that they share common domains. Understanding these domains and their function, it will be possible to devise plant protection strategies using these genes that provides broad spectrum resistance against a variety of pathogens.

*Nematode resistance*: A variety of approaches have been utilised to protect against damage from nematodes. Regulatory regions have been identified which are uniquely engaged when the nematode burrows into the root; simple physical damage of the root is not sufficient to turn on the promoter. This regulatory region can then be hooked to nematicidal proteins, which can then be delivered to the nematode at its precise site of attack. Such nematicidal proteins include several different proteinase inhibitors and lectins, which act in some cases synergitically to interfere with the sexual development of the nematode.

In addition, recently a natural resistance gene was isolated from tomato and found to provide protection in engineered plants to not only nematodes but also aphids, which have an entirely different mode of attack and damage pathway. It was found, after comparison of the gene to other natural resistance genes (fungal, viral and bacterial) from other species, that the nematode resistance gene shared common domains with these other genes, again pointing to the potential universality of these mechanisms and opening the door for creating some engineered proteins with widespread efficacy.

### Harvest and Post-harvest Characteristics

Another arena in which we have seen product enter the marketplace and in which much research is being conducted relates to harvest and post-harvest characteristics. Two different approaches to delayed ripening have been commercialised. The approach of the Flavr Savr tomato utilised a technique allowing growers and producers extra time to pick and deliver tomatoes to the marketplace by interfering with polygalacturonase synthesis, an enzyme involved in pectin deposition. In this case the ripening process was delayed by several days but was not halted.

In another approach to interfering with the ripening process, researchers stopped the ripening process completely by interfering with the production of ethylene, which is essential to ripening. By this process it is possible to leave a tomato on the vine for up to 100 days past the time when it would ordinarily be ripe and, once it is picked, to ripen the tomato on demand by exposure to ethylene.

Another post-harvest characteristic, which is the focus of genetic engineering technologies is the achievement of more uniform ripening. This was first accomplished with the cherry tomato, but the basic approach is also being attempted with processing tomatoes in order to achieve less waste and aid harvesting. This approach involves the manipulation of genes involved in slowing the degradation of ethylene.

### Agronomic Traits

*Nitrogen fixation and utilisation*: Fundamental work on the process of nitrogen fixation has been pursued for decades. With the advent of molecular techniques, pace of that research has increased and has broadened to include fundamental studies on the nature of the plant-microbe interaction necessary for nodule formation and nitrogen fixation. Work on the bacterium itself has resulted in engineered Rhizobium strains that have increased rates of nitrogen fixation in limited field trials, but those strains have not yet been released for widespread use by the EPA.

Studies focused on understanding and manipulating the host range of nitrogen-fixing bacteria are also underway with the expectation of being able to extend the range of crop species that can have productive commensal relationships with bacterial nitrogen fixers. In addition to looking at improving the efficiency of nitrogen fixation, fundamental work aimed at understanding nitrogen fertiliser use efficiency is also underway. This work hopes to lead to engineered plants, e.g. wheat and corn, that more efficiently use applied fertilisers.

### Male Sterility

The development of hybrid varieties that result in increased yields of crop species has been a tedious process, requiring in many cases the manual detasseling of the male parent to prevent self-fertilisation. Large scale production of hybrid seed, however, required more efficient mechanisms to suppress pollen production. Classical geneticists have identified naturally occurring male-sterile lines of some crops that were used for commercial seed production. Such strategies also required the identification of "restorer lines" that could restore male fertility by counteracting the genes that cause male-sterility. This is more easily accomplished with some crops than with others. With the advent of genetic engineering, it became possible to develop strategies to create male sterile plants. These strategies made use of the fact that one could identify regulatory sequences that would lead to very specific tissue expression.

In one strategy a promoter was identified that only drove expression in the inner layer of the pollen sac, the tapetum. Linked to this promoter was a bacterial enzyme, which degrades RNA, thereby interrupting the development pollen grains. A restorer of fertility was also identified; this protein is a chemical coupler of the RNA-degrading enzyme; the tight coupling of these two proteins inactivates the RNA-degrading enzyme, thus restoring fertility. This approach was used in canola first and is now being attempted in a variety of different crop and vegetable species.

### Terminator Technology

This approach is a relatively new one patented by a small cotton company, Delta and Pine Land, and the USDA; the Monsanto Corporation has made a bid to buy DP&L, but that deal is still awaiting approval by the Department of Justice. The fundamental idea with this technology is to prevent a plant from reproducing itself by interfering with seed development, thus insuring that the user of the seed cannot replant a field without repurchasing the seed, thereby protecting the company's investment in developing the variety.

The system works through the interplay of three introduced genes. One of these genes leads to the production of a toxin that kills the seed in the final stage of development. This gene is initially interrupted by a stretch of DNA between the toxin gene itself and the regulatory region or promoter that is responsible for its synthesis. This gene configuration can then be transplanted or crossed with another genetically engineered plant that produces a scissor-like enzyme, capable of removing the segment in the toxin gene, thereby activating it and causing the seed to die. The third component of the system is a repressor gene that prevents the synthesis of the scissor-like enzyme. If all three components are present, the terminator technology is dormant and the seed is fertile.

DP&L is currently pursuing this technique in cotton and wheat and expect these products to be on the market by 2005. With the entry of Monsanto into the equation, application of the technology could accelerate into many more crops and the advent into the marketplace could be much more rapid. There has been much furor and debate about this technology since obviously it can be viewed as an interference with farmer's rights and insures that farmers return every year to purchase a new lot of seeds. This has led a large, international, not-for-profit agricultural research organisation, the Consultative Group on International Agricultural Research, or CGIAR, to consider banning the use of the technology in crop development programmes.

Another concern held by environmentalist and certain agriculturalists is that certain traits, like herbicide tolerance and certain other traits, will escape from the crop plant and enter into wild species located nearby, thereby creating situation where it is difficult to control the wild species or it has some other new, adverse characteristic. This issue has been raised with, for example, sorghum and Johnson grass, oat and wild oat, rice and red rice, canola and its wild Brassica relatives. The use of such a technology for engineering plants that have wild relatives nearby is a prudent approach to preventing the dissemination of genes to wild populations.

### Value-added Products

There is another area that does not involve plant growth regulators but is being intensely pursued by both the public and private sectors. This relates to the improvement of existing plant products, ones that already exist but for which new or improved functions can be identified that increase the value of the product. An example would be specialty oils from rape or soybean, which are designed to have improved heat stability. Another example would be the genetic engineering of sugarbeet to produce fructans, a sweet, no-calorie sugar substitute.

An intriguing application is in the area of functional foods or neutriceuticals. This line of foods is based on the fact that we can now identify in many cases natural products in foods that can either cause adverse human condition or prevent disease. Examples of those products focused on factors in foods that cause disease or undesirable pharmacological effects include ameliorating the allergenic potential of foods, such as wheat, milk and peanut or removing caffeine from coffee. In the category of natural products that prevent disease, there are many examples.

Transgenic potatoes are being created which produce high quantities of GAD, an enzyme believed to be involved in the prevention of type II juvenile diabetes. Researchers have identified an element in grapes that increases blood flow in blood vessels; grapes could be engineered to increase the level of such an element.

# Bibliography


Davis, T. D. and Haissig, B.E., "Chenical control of adventitious root formation in cuttings", *Bulletin of Plant Growth Regulators of Society American*, 1990.

________________ , and Sankhla, N., "Effect of shoot growth retardants and inhibitors on adventitious rooting", In: Davis,T. D.; Haissig, B. E. and Sankhla, N. (Eds.), *Adventitious root formation in cuttings*, Portland: Dioscoride Press, 1988.

Glover, C.R., "Pix and Prep use in cotton", New Mexico State University, cooperative Extension Service, Guide A-213, 1992

Hartmann, H. T.; Kester, D. E.; Davis Jr, F. T. and Geneve, R. L., "Plant propagation: principles and practices", New Jersey: Prentice Hall, 2002.

Leopold, A. C., *Plant growth and development*, New York: Mc Graw-Hill, 1964.

Nelson, C. D.; Linghai, Z. and Hamaker, J. M., "Propagation of loblolly, slash, and Longleaf pine from needle fascicles", *Tree Plant Notes*, 1992.

Nelson, J.M., F.S. Nakayama, H.M. Flint, R.L. Garcia, and C.L. Hart, "Methanol treatments on Pima and upland cotton. Cotton", *A College of Agriculture Report*, The University of Arizona, Tucson, AZ. Series: Oklahoma State University, 1995.

Nelson, N., "A photometric adaptation of the Somogy method for determination of glucose", *Journal of Biology Chemistry*, 1944.

Oosterhuis, D.M., and D. Zhao., "Physiological Effects of PGR IV on the Growth and Yield of Cotton", *Beltwide Cotton Conference Proceedings*, 1993.

Pan, R. and Zhao, Z., "Synergistic effects of plant growth retardants and IBA on the formation of adventitious roots in hypocotyl cuttings of mungbean", *Plant Growth Regulators*, 1994.

Pandey, D. and Pathak, R. K., "Effect of Rootstocks, IBA and Phenolic Compounds on the Rooting of Apple Cuttings", *Propagation of Horticulture*, 1981.

Rademacher, W.; Jung, J.; Grabe, J. E.; Schwendem, L., "On the mode of action of tetcyclasis and triazole growth retardants", *Br Plant Growth Monograph*, 1984.

Robertson, W.C., and J.T. Cothren. 1993. "PGR IV Effects on Boll Distribution, Lint Yield, and Fiber Quality", *Beltwide Cotton Conference Proceedings*, 1993.

Silvertooth, J.C., S.W. Stedman, R.E. Cluff, and E.R. Norton, "Cotton Defoliation Evaluations", *Cotton, A College of Agriculture Report*, The University of Arizona, Tucson, AZ, 1993.

# Index